# FE/EIT

## Sample Examinations

### Second Edition

## Michael R. Lindeburg, PE

Professional Publications, Inc. • Belmont, CA

# How to Locate and Report Errata for This Book

At Professional Publications, we do our best to bring you error-free books. But when errors do occur, we want to make sure you can view corrections and report any potential errors you find, so the errors cause as little confusion as possible.

A current list of known errata and other updates for this book is available on the PPI website at **www.ppi2pass.com/errata**. We update the errata page as often as necessary, so check in regularly. You will also find instructions for submitting suspected errata. We are grateful to every reader who takes the time to help us improve the quality of our books by pointing out an error.

**FE/EIT Sample Examinations**
**Second Edition**

Current printing of this edition:   2

**Printing History**

| edition number | printing number | update |
|---|---|---|
| 1 | 3 | Minor corrections. |
| 2 | 1 | New edition. |
| 2 | 2 | Minor corrections. |

Printed in the United States of America

PPI
1250 Fifth Avenue, Belmont, CA 94002
(650) 593-9119
www.ppi2pass.com

**Library of Congress Cataloging-in-Publication Data**
Lindeburg, Michael R.
    FE/EIT sample examinations / Michael R. Lindeburg. -- 2nd ed.
      p. cm.
  ISBN-13: 978-1-59126-074-5
  ISBN-10: 1-59126-074-4
  1. Engineering--United States--Examinations, questions, etc. 2. Civil
engineering--United States--Examinations, questions, etc. 3. Engineering--Problems,
 exercises, etc. 4. Engineers--Certification--United States. I. Title.

TA159.L573 2006
620.0076--dc22

2006041812

# Table of Contents

# Preface

I can still remember sitting down and writing my first FE sample exam in 1975. (In those days, it was called the "EIT exam.") I had been out of college less than five years when I found myself teaching at a local college in California. Soon, I was providing sample exams to review courses in local companies (Lockheed, Bechtel, Memorex, etc.) as well. My initial one-off sample exams were pretty sophomoric. Luckily, by the time I got around to writing *EIT Sample Examinations*, the now-out-of-print predecessor to this publication, I had figured out that examinees considered their exam preparations to be a serious matter. Out went the questions pertaining to widgets, furlongs per fortnight, and companies such as Sayure Prairie Airlines and Where's the Beef Ranch.

Since those early days, this publication has changed its name, replaced almost all of the questions, doubled its value (i.e., provided two sample exams in place of just one), and bobbed and weaved through numerous changes in exam format. This second edition incorporates the changes in exam format announced in October 2005 by NCEES for examinations starting in 2006. The most significant changes to the FE exam with a general afternoon emphasis are the addition of biology, increased emphasis on probability and statistics, elimination of three-phase electricity, decreased emphasis on electronics and some computer topics, and a complete redistribution of numbers of questions in each topic.

As with any sample exam that I write, there is always the Author's Caveat: Use this exam as an evaluation, not as a preparation outline. This exam, like the real FE exam, contains its share of red herrings (i.e., questions requiring far more time than you have per question, questions answerable only by engineers who majored in a particular discipline, and questions that cover obscure subjects). By including these types of questions, I'm not suggesting that you become proficient in their underlying knowledge bases. Your preparation should be guided by the *FE Review Manual* or your review course, not this examination. If you use this examination to guide your study, you'll end up learning a lot about subjects that you won't see in the real exam, without protecting yourself from similar red herrings from another universe.

Nobody knows better than an author that multiple stages of editing, calculation checking, and proofreading are worth doing but still inadequate to produce an error-free book. If you think you've found a mistake, I hope you'll let me know about it. It's pretty simple to use PPI's online errata submission form to submit what you find.

So, off you go. I wrote this book to help you. Now, it's up to you.

Michael R. Lindeburg, PE

# Acknowledgments

This new edition contains new material written by Rhandi Gallegos, PE (several topics, though primarily fluid dynamics), and Jamie Rana, PE (engineering mechanics and strength of materials). Rhandi contributed a couple dozen questions in areas that I requested. Jamie also technically reviewed some of the new material authored by others. PPI's own Sarah Hubbard, Editorial Department Manager, wrote about two dozen new problems and recast some existing problems before taking a maternity leave of absence. One of our editors, Scott Marley, brought his mathematical knowledge and puzzle orientation to bear as he wrote new probability, statistics, and economics problems. Mitch Bakos (BSEE, FE) from PPI's IT department had a hand in resolving vagaries in several electrical problems.

The project of managing the publication of this new edition was the responsibility of PPI Project Editor Dennis Rowcliffe. He monitored everyone's work, and he sounded the alarms when the book wasn't going in the correct direction. He worked diligently to bring out a book with just the right number of questions from each topic, each at just the right difficulty level. Late in the game, when the book had fallen behind schedule and still had the wrong mix of problems per discipline, he helped persuade Rhandi and Jamie to do more in their particular specialties, and he, Sarah, and Scott got things heading in the right direction. Some of the new biology and environmental problems are based on similar introductory material in my Environmental Engineering Reference Manual, and Dennis was instrumental in adapting that material for use in this publication.

In addition to Dennis, my daughter, Jenny, proofread the final rounds. She cut her teeth on this book as an editor, checking for consistency and accuracy in the material. She brought a fresh perspective to the task. I don't know if she ever expected that the Editor-in-Chief reviewing her work would be her father, but saying that I'm pretty tickled about the arrangement is an understatement.

Cathy Schrott managed all of the production (i.e., typesetting and illustrating) activities, though she would rather have had nothing to do with the mess she was given. She took a manuscript having six points of origin and imposed a structure on the project that worked. Incidentally, Cathy was the lead typesetter of the original edition of this book, though that was in a parallel universe in another lifetime. Miriam Hanes did a masterful job of typesetting and paginating this edition of the book. She also made a lot of corrections and improvements on the fly, as did all of PPI's upstairs crew, who helped each other. Kate Hayes typeset the handwritten material that I and others submitted (in violation of our contracts, I am sure). Tom Bergstrom rendered the new illustrations required by the new questions, and he also updated the existing illustrations to PPI's Style Guide, which simply refuses to remain static over the lifetime of any given title. Amy Schwertman again proved that she's in the right place doing what she does best by designing and rendering the cover so that it now, clearly, is part of PPI's line of FE books.

Numerous other people, having contributed to previous editions and printings, now qualify for the title of legacy contributors. They include Joseph Sheeley, then a PhD candidate from the University of California, Berkeley, Steve Van Wyk, PE, who instructed at Olympic College in Washington as well as teaching his own review courses, and Dina Hadi, an engineering intern working at PPI while finishing up her Mechanical Engineering degree at the University of Santa Clara. Their contributions are not forgotten.

What a humbling collection of names is the above list. Thank you, all!

Michael R. Lindeburg, PE

# Introduction

**I know the format of the FE exam, but what does it look like?**
That is a question I get frequently, and such is the question that *FE/EIT Sample Examinations* seeks to answer.

**What kinds of questions are there?**
Lots of kinds.

**How difficult are the questions?**
Generally, not very.

**How obtuse are they?**
It varies from question to question and depends on you.

**Are the four answer choices so close that you have to carry all calculations out to six significant digits?**
No.

**Are the choices tricky, so if you forget to convert centimeters to meters you actually find a logical wrong answer?**
Yes.

*FE/EIT Sample Examinations* is the paradigm that will give you a feel for the answers to these questions.

There are two full-length (i.e., eight-hour) sample exams in this publication, along with full solutions. This book stands on its own, with no specific references to other books in the solutions. However, the symbols are the same as those commonly used, the equations are given symbolically before the values are entered, and all calculations include units. It should be a simple matter to follow the solution technique.

As with any representative exam, these two exams can be used to evaluate your preparedness, but they should not be used to guide your preparation. There is no guarantee that any problem type in this publication ever has or ever will appear on any FE examination. Some of the problems will be too simple, others will be too difficult, and some will come from far out in left field. Basing your preparation on these specific problems rather than preparing for a wide range of problem types will only get you in trouble.

Michael R. Lindeburg, PE

# How To Use This Book

This book contains two sample exams, so there are only a few ways that you can use it. Most people will work through every problem, basically using this book as a collection of solved problems. At the other extreme, some people will run out of time and won't crack the book at all. A few people will actually use the book as I intended: by taking the exams under realistic, timed conditions.

But, to me, the main issue is not *how* you use this sample exam, but *when* you use it.

Though I tried to include realistic exam problems, I did not write this book intending it to be a diagnostic tool guiding your preparation. You shouldn't take this sample exam and then design your review around what you didn't know. If you take this exam and don't do well on a particular problem, I wouldn't want you to spend the next three months preparing for that type of problem. The tried-and-true method of exam preparation is a systematic, thorough, and complete approach based on long-term exam trends. That's what my *FE Review Manual* and *Engineer-in-Training Reference Manual* are for.

The value of a sample exam is not in its ability to guide your preparation. Rather, the value is in giving you an opportunity to bring together all of your knowledge and to practice your test-taking skills. The three most important skills are (1) selecting the right problems, (2) becoming familiar with the content and organization of the NCEES Handbook, and (3) managing your time. I intended that these sample exams be taken within a few weeks of your actual exam. That's the only time that you will be able to focus on test-taking skills without the distraction of rusty recall.

You'll need to set aside an entire day in order to take each sample exam the way I intended. I know using up another day is asking a lot from you. But if you start early enough and study diligently, by the time the actual exam rolls around, you will probably be weak in only one area: familiarity with the nature of the exam.

In athletics, coaches usually prefer a home-field advantage. Athletes who sleep in their own beds, dress out of their own lockers, and play on their own fields play better. Well, examinees who have "seen it before" via a sample exam have a psychological advantage, as well. This publication was written to give you that edge.

Good luck!

# Instructions

This section of the exam consists of 120 problems, each worth 1 point. You will have four hours in which to work this section. Your score will be determined by the number of problems that you solve correctly. No points will be deducted for incorrect answers. It is to your best advantage to try to answer every question.

When permission has been given by your proctor, break the seal on the Examination Booklet and remove the Answer Sheet. Write your name immediately in the space indicated. Check that all pages are present and legible. If any part of this Booklet is missing, your proctor will issue you a new Booklet.

All solutions must be entered on the Answer Sheet. No credit will be given for answers appearing only in the Examination Booklet. Mark your answers with the pencil provided. Do not use pen. Marks must be dark and must completely fill the bubble. Record only one answer per problem; if you mark more than one answer, you will not receive credit for the problem. If you change an answer, be sure the old bubble is erased completely; incomplete erasures may be read as intended answers.

If you finish early, check your work and make sure you have correctly followed all instructions. After checking your answers, you may turn in your Examination Booklet and Answer Sheet and leave the examination room. Once you leave, you will not be permitted to return to work on your solutions.

Do not work any problems from the Afternoon Section of the exam during the first four hours of this exam.

WAIT FOR PERMISSION TO BEGIN.

Name: _____
   Last   First   Middle
              Initial

## FUNDAMENTALS OF ENGINEERING SAMPLE EXAMINATION

### EXAM 1
### MORNING SECTION

#### Subject Breakdown

The major subject areas and their corresponding problem numbers are listed below.

| | |
|---|---|
| Mathematics | 1–18 |
| Engineering Probability and Statistics | 19–26 |
| Chemistry | 27–36 |
| Computers | 37–45 |
| Ethics and Business Practices | 46–54 |
| Engineering Economics | 55–63 |
| Engineering Mechanics (Statics and Dynamics) | 64–75 |
| Strength of Materials | 76–84 |
| Material Properties | 85–92 |
| Fluid Mechanics | 93–101 |
| Electricity and Magnetism | 102–111 |
| Thermodynamics | 112–120 |

# Fundamentals of Engineering Sample Examination

## Morning Section

Name: _____

| | | | |
|---|---|---|---|
| 1. Ⓐ Ⓑ Ⓒ Ⓓ | 31. Ⓐ Ⓑ Ⓒ Ⓓ | 61. Ⓐ Ⓑ Ⓒ Ⓓ | 91. Ⓐ Ⓑ Ⓒ Ⓓ |
| 2. Ⓐ Ⓑ Ⓒ Ⓓ | 32. Ⓐ Ⓑ Ⓒ Ⓓ | 62. Ⓐ Ⓑ Ⓒ Ⓓ | 92. Ⓐ Ⓑ Ⓒ Ⓓ |
| 3. Ⓐ Ⓑ Ⓒ Ⓓ | 33. Ⓐ Ⓑ Ⓒ Ⓓ | 63. Ⓐ Ⓑ Ⓒ Ⓓ | 93. Ⓐ Ⓑ Ⓒ Ⓓ |
| 4. Ⓐ Ⓑ Ⓒ Ⓓ | 34. Ⓐ Ⓑ Ⓒ Ⓓ | 64. Ⓐ Ⓑ Ⓒ Ⓓ | 94. Ⓐ Ⓑ Ⓒ Ⓓ |
| 5. Ⓐ Ⓑ Ⓒ Ⓓ | 35. Ⓐ Ⓑ Ⓒ Ⓓ | 65. Ⓐ Ⓑ Ⓒ Ⓓ | 95. Ⓐ Ⓑ Ⓒ Ⓓ |
| 6. Ⓐ Ⓑ Ⓒ Ⓓ | 36. Ⓐ Ⓑ Ⓒ Ⓓ | 66. Ⓐ Ⓑ Ⓒ Ⓓ | 96. Ⓐ Ⓑ Ⓒ Ⓓ |
| 7. Ⓐ Ⓑ Ⓒ Ⓓ | 37. Ⓐ Ⓑ Ⓒ Ⓓ | 67. Ⓐ Ⓑ Ⓒ Ⓓ | 97. Ⓐ Ⓑ Ⓒ Ⓓ |
| 8. Ⓐ Ⓑ Ⓒ Ⓓ | 38. Ⓐ Ⓑ Ⓒ Ⓓ | 68. Ⓐ Ⓑ Ⓒ Ⓓ | 98. Ⓐ Ⓑ Ⓒ Ⓓ |
| 9. Ⓐ Ⓑ Ⓒ Ⓓ | 39. Ⓐ Ⓑ Ⓒ Ⓓ | 69. Ⓐ Ⓑ Ⓒ Ⓓ | 99. Ⓐ Ⓑ Ⓒ Ⓓ |
| 10. Ⓐ Ⓑ Ⓒ Ⓓ | 40. Ⓐ Ⓑ Ⓒ Ⓓ | 70. Ⓐ Ⓑ Ⓒ Ⓓ | 100. Ⓐ Ⓑ Ⓒ Ⓓ |
| 11. Ⓐ Ⓑ Ⓒ Ⓓ | 41. Ⓐ Ⓑ Ⓒ Ⓓ | 71. Ⓐ Ⓑ Ⓒ Ⓓ | 101. Ⓐ Ⓑ Ⓒ Ⓓ |
| 12. Ⓐ Ⓑ Ⓒ Ⓓ | 42. Ⓐ Ⓑ Ⓒ Ⓓ | 72. Ⓐ Ⓑ Ⓒ Ⓓ | 102. Ⓐ Ⓑ Ⓒ Ⓓ |
| 13. Ⓐ Ⓑ Ⓒ Ⓓ | 43. Ⓐ Ⓑ Ⓒ Ⓓ | 73. Ⓐ Ⓑ Ⓒ Ⓓ | 103. Ⓐ Ⓑ Ⓒ Ⓓ |
| 14. Ⓐ Ⓑ Ⓒ Ⓓ | 44. Ⓐ Ⓑ Ⓒ Ⓓ | 74. Ⓐ Ⓑ Ⓒ Ⓓ | 104. Ⓐ Ⓑ Ⓒ Ⓓ |
| 15. Ⓐ Ⓑ Ⓒ Ⓓ | 45. Ⓐ Ⓑ Ⓒ Ⓓ | 75. Ⓐ Ⓑ Ⓒ Ⓓ | 105. Ⓐ Ⓑ Ⓒ Ⓓ |
| 16. Ⓐ Ⓑ Ⓒ Ⓓ | 46. Ⓐ Ⓑ Ⓒ Ⓓ | 76. Ⓐ Ⓑ Ⓒ Ⓓ | 106. Ⓐ Ⓑ Ⓒ Ⓓ |
| 17. Ⓐ Ⓑ Ⓒ Ⓓ | 47. Ⓐ Ⓑ Ⓒ Ⓓ | 77. Ⓐ Ⓑ Ⓒ Ⓓ | 107. Ⓐ Ⓑ Ⓒ Ⓓ |
| 18. Ⓐ Ⓑ Ⓒ Ⓓ | 48. Ⓐ Ⓑ Ⓒ Ⓓ | 78. Ⓐ Ⓑ Ⓒ Ⓓ | 108. Ⓐ Ⓑ Ⓒ Ⓓ |
| 19. Ⓐ Ⓑ Ⓒ Ⓓ | 49. Ⓐ Ⓑ Ⓒ Ⓓ | 79. Ⓐ Ⓑ Ⓒ Ⓓ | 109. Ⓐ Ⓑ Ⓒ Ⓓ |
| 20. Ⓐ Ⓑ Ⓒ Ⓓ | 50. Ⓐ Ⓑ Ⓒ Ⓓ | 80. Ⓐ Ⓑ Ⓒ Ⓓ | 110. Ⓐ Ⓑ Ⓒ Ⓓ |
| 21. Ⓐ Ⓑ Ⓒ Ⓓ | 51. Ⓐ Ⓑ Ⓒ Ⓓ | 81. Ⓐ Ⓑ Ⓒ Ⓓ | 111. Ⓐ Ⓑ Ⓒ Ⓓ |
| 22. Ⓐ Ⓑ Ⓒ Ⓓ | 52. Ⓐ Ⓑ Ⓒ Ⓓ | 82. Ⓐ Ⓑ Ⓒ Ⓓ | 112. Ⓐ Ⓑ Ⓒ Ⓓ |
| 23. Ⓐ Ⓑ Ⓒ Ⓓ | 53. Ⓐ Ⓑ Ⓒ Ⓓ | 83. Ⓐ Ⓑ Ⓒ Ⓓ | 113. Ⓐ Ⓑ Ⓒ Ⓓ |
| 24. Ⓐ Ⓑ Ⓒ Ⓓ | 54. Ⓐ Ⓑ Ⓒ Ⓓ | 84. Ⓐ Ⓑ Ⓒ Ⓓ | 114. Ⓐ Ⓑ Ⓒ Ⓓ |
| 25. Ⓐ Ⓑ Ⓒ Ⓓ | 55. Ⓐ Ⓑ Ⓒ Ⓓ | 85. Ⓐ Ⓑ Ⓒ Ⓓ | 115. Ⓐ Ⓑ Ⓒ Ⓓ |
| 26. Ⓐ Ⓑ Ⓒ Ⓓ | 56. Ⓐ Ⓑ Ⓒ Ⓓ | 86. Ⓐ Ⓑ Ⓒ Ⓓ | 116. Ⓐ Ⓑ Ⓒ Ⓓ |
| 27. Ⓐ Ⓑ Ⓒ Ⓓ | 57. Ⓐ Ⓑ Ⓒ Ⓓ | 87. Ⓐ Ⓑ Ⓒ Ⓓ | 117. Ⓐ Ⓑ Ⓒ Ⓓ |
| 28. Ⓐ Ⓑ Ⓒ Ⓓ | 58. Ⓐ Ⓑ Ⓒ Ⓓ | 88. Ⓐ Ⓑ Ⓒ Ⓓ | 118. Ⓐ Ⓑ Ⓒ Ⓓ |
| 29. Ⓐ Ⓑ Ⓒ Ⓓ | 59. Ⓐ Ⓑ Ⓒ Ⓓ | 89. Ⓐ Ⓑ Ⓒ Ⓓ | 119. Ⓐ Ⓑ Ⓒ Ⓓ |
| 30. Ⓐ Ⓑ Ⓒ Ⓓ | 60. Ⓐ Ⓑ Ⓒ Ⓓ | 90. Ⓐ Ⓑ Ⓒ Ⓓ | 120. Ⓐ Ⓑ Ⓒ Ⓓ |

**1.** The half-life of a radioactive isotope is 4.3 days. Most nearly, how long will it take to reduce the original amount to 1%?

(A) 3.2 days
(B) 16 days
(C) 29 days
(D) 40 days

**2.** What is the determinant of the following $2 \times 2$ matrix?

$$\begin{bmatrix} 1 & 4 \\ 3 & 2 \end{bmatrix}$$

(A) $-10$
(B) $-5$
(C) 9
(D) 10

**3.** What is most nearly the value of the following limit?

$$\lim_{x \to 3} \frac{x^2 - 9}{x - 3}$$

(A) $-6$
(B) 1
(C) 6
(D) $\infty$

**4.** Given the vector $\mathbf{V} = \mathbf{i} + 2\mathbf{j} + \mathbf{k}$, what is most nearly the angle between $\mathbf{V}$ and the $x$-axis?

(A) $22°$
(B) $24°$
(C) $66°$
(D) $80°$

**5.** Evaluate the following definite integral.

$$\int_{2}^{\infty} \frac{1}{x^2}\, dx$$

(A) 1/24
(B) 1/8
(C) 1/2
(D) 2

**6.** Which is a true statement about the two vectors?

$$\mathbf{V}_1 = \mathbf{i} + 2\mathbf{j} + \mathbf{k}$$
$$\mathbf{V}_2 = \mathbf{i} + 3\mathbf{j} - 7\mathbf{k}$$

(A) Both vectors pass through $(0, -1, 6)$.
(B) The vectors are parallel.
(C) The vectors are orthogonal.
(D) The angle between the vectors is $17.4°$.

**7.** What is most nearly the area bounded by $y = 0$, $y = e^x$, $x = 0$, and $x = 1$?

(A) 1.4
(B) 1.7
(C) 2.7
(D) 3.4

**8.** A function of $x$ is given below. Which $(x, y)$ point is a relative maximum or minimum?

$$y = \tfrac{1}{4}x^4 - 1.5x^2 + 2x + 5$$

(A) $(-2, -1)$
(B) $(-2, -2)$
(C) $(2, -2)$
(D) $(-1, -1.75)$

**9.** The slope of a line is $^1/_2$. The slope of a second line is $-^2/_3$. The lines intercept at the point $(3, 1)$. What is most nearly the acute angle between the lines?

(A) $27°$
(B) $50°$
(C) $60°$
(D) $80°$

**10.** A function is given. What value of $x$ maximizes $y$?

$$y^2 + y + x^2 - 2x = 5$$

(A) $-1$
(B) 1/2
(C) 1
(D) 5

**11.** An elephant is chained to a corner of a 30 m $\times$ 35 m building. If the chain is 40 m long and the elephant can reach 1 m farther than the chain length, what is the maximum area the elephant can cover?

(A) $3870 \text{ m}^2$
(B) $3960 \text{ m}^2$
(C) $3970 \text{ m}^2$
(D) $4080 \text{ m}^2$

**12.** What is a point of inflection for the following equation?

$$y = 9x^3 + x^2 - 15x + 32$$

(A) $(-27, -176{,}000)$
(B) $(-21.2, -84{,}400)$
(C) $(-0.037, 32.6)$
(D) $(19.2, 63{,}300)$

**13.** Assume that three force vectors are applied at a single point.

$$\mathbf{F}_1 = \mathbf{i} + 3\mathbf{j} + 4\mathbf{k}$$
$$\mathbf{F}_2 = 2\mathbf{i} + 7\mathbf{j} - \mathbf{k}$$
$$\mathbf{F}_3 = -\mathbf{i} + 4\mathbf{j} + 2\mathbf{k}$$

What is most nearly the magnitude of the resultant force vector, $\mathbf{R}$?

(A) 13
(B) 14
(C) 15
(D) 16

**14.** What is the solution to the differential equation if $x = 1$ at $t = 0$, and $dx/dt = 0$ at $t = 0$?

$$\frac{1}{2}\frac{d^2x}{dt^2} + 4\frac{dx}{dt} + 8x = 5$$

(A) $e^{-4t} + 4te^{-4t}$
(B) $\frac{3}{8}e^{-2t}(\cos 2t + \sin 2t) + \frac{5}{8}$
(C) $e^{-4t} + 4te^{-4t} + \frac{5}{8}$
(D) $\frac{3}{8}e^{-4t} + \frac{3}{2}te^{-4t} + \frac{5}{8}$

**15.** What is the smallest positive value of $y$ on the curve $y = 7x^2 - 3x + 8$?

(A) 3/14
(B) 3/7
(C) 14/3
(D) 215/28

**16.** Which of the following is equivalent to $\sin 2\theta$?

(A) $2\sin\theta\cos\theta$
(B) $\cos^2\theta - \sin^2\theta$
(C) $\sin\theta\cos\theta$
(D) $\dfrac{1 - \cos 2\theta}{2}$

**17.** If $i \equiv \sqrt{-1}$, what is the value of $(i)^i$?

(A) $i^2$
(B) $e^{2i}$
(C) $-1$
(D) $e^{-\frac{\pi}{2}}$

**18.** What is most nearly the volume of the object created when the area bounded by $y = 0$, $x = 0$, and $y = \sqrt{4 - x^2}$ is rotated about the $y$-axis?

(A) 3.1
(B) 8.4
(C) 17
(D) 34

**19.** Five fair coins are each flipped once. What is the probability that at least two of the coins will show heads?

(A) 0.19
(B) 0.80
(C) 0.81
(D) 0.84

**20.** What is most nearly the standard deviation of 1, 4, and 7?

(A) 2.5
(B) 3.0
(C) 5.7
(D) 6.0

**21.** What is most nearly the probability of picking an orange ball out of a bag containing seven orange balls, eight green balls, and two white balls?

(A) 0.059
(B) 0.15
(C) 0.24
(D) 0.41

**22.** Five dice are thrown. The probability of rollling at least one six is most nearly

(A) 0.50
(B) 0.52
(C) 0.60
(D) 0.67

**23.** What is most nearly the sample standard deviation for the data set $\{2.0, 7.0, 9.0, 12, 34\}$?

(A) 11
(B) 12
(C) 13
(D) 17

**24.** On average, a piece of machinery jams five times a week. Assuming a Poisson distribution for the jams, the probability that the machine will jam exactly three times in a given week is most nearly

(A)  0.0033
(B)  0.14
(C)  0.33
(D)  1.40

**25.**  The least-squares method is used to plot a straight line through the data points (2,2), (4,7), (1,11), and (5,9). The slope of the line is most nearly

(A)  $-8$
(B)  $-4$
(C)  0
(D)  3.5

**26.**  The method of determining the graph of a curve that best approximates a given collection of points is called

(A)  integration by parts
(B)  curve fitting
(C)  nonlinear interpolation
(D)  hypothesis testing

**27.**  Which of the following is the formula for acetic acid?

(A)  COOH
(B)  $CH_2COOH$
(C)  $CH_3CH_2COOH$
(D)  $CH_3COOH$

**28.**  What is oxidized and what is reduced in the following reaction?

$$Zn + H_2SO_4 \rightarrow ZnSO_4 + H_2$$

(A)  Zinc is oxidized only.
(B)  Zinc is reduced only.
(C)  Zinc is oxidized and hydrogen is reduced.
(D)  Zinc is reduced and hydrogen is oxidized.

**29.**  How many moles of NaOH will be neutralized by 1 mole of $H_3PO_4$?

(A)  1/3
(B)  1
(C)  2
(D)  3

**30.**  An anode is

(A)  the endpoint of a directed network
(B)  an electrode at which oxidation occurs
(C)  an electrode at which reduction occurs
(D)  the electrode to which the cation would be attracted during an electrolytic reaction

**31.**  Beryllium, magnesium, and calcium all belong to which elemental grouping?

(A)  noble elements
(B)  halogens
(C)  alkali metals
(D)  alkaline earth metals

**32.**  If 1 mol of gaseous chlorine combines with 1 mol of calcium to form calcium chloride ($CaCl_2$), which of the following statements is true?

(A)  Calcium has a valence of $+1$.
(B)  1 lbm of chlorine will combine with 1 lbm of calcium.
(C)  Chlorine has a valence of $-2$.
(D)  Calcium has a valence of $+2$.

**33.**  Balance the following equation.

$$\_\_ PbO_2 + \_\_ H_2SO_4 + \_\_ Pb \rightleftharpoons \_\_ H_2O + \_\_ PbSO_4$$

(A)  $3PbO_2 + 6H_2SO_4 + 3Pb \rightleftharpoons 6H_2O + 5PbSO_4$
(B)  $3PbO_2 + 6H_2SO_4 + Pb \rightleftharpoons 6H_2O + 4PbSO_4$
(C)  $2PbO_2 + H_2SO_4 + 2Pb \rightleftharpoons 4H_2O + 4PbSO_4$
(D)  $PbO_2 + 2H_2SO_4 + Pb \rightleftharpoons 2H_2O + 2PbSO_4$

**34.**  What is the most likely formula of a compound with the following gravimetric analysis?

oxygen:  13.7%
carbon:  20.5%
hydrogen:  5.1%
chlorine:  60.7%

(A)  $CH_3OCl$
(B)  $C_2H_6OCl$
(C)  $C_2H_6OCl_2$
(D)  $CH_6O_2Cl$

**35.**  How many moles of hydrochloric acid will be required to neutralize 1 mol of sodium hydroxide?

(A)  0.3 mol
(B)  0.5 mol
(C)  1.0 mol
(D)  1.5 mol

**36.**  What is most nearly the mass of 1 atom of carbon-12?

(A)  $5.0 \times 10^{-24}$ g
(B)  $2.0 \times 10^{-23}$ g
(C)  $4.0 \times 10^{-23}$ g
(D)  $7.6 \times 10^{-23}$ g

**37.** A system is found to have the following transfer function.

$$H(s) = \frac{7+s}{2s^2 + 14s - 36}$$

$s$ is the frequency in Hz. For what value of $s$ does the system become unstable?

(A)  −7 Hz
(B)  2 Hz
(C)  4 Hz
(D)  9 Hz

Problems 38–43 are based on the following illustration. All values are in non-dimensionalized units.

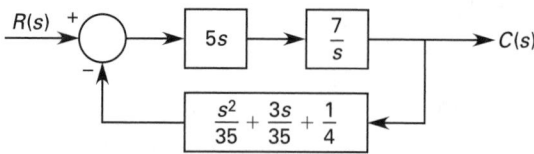

**38.** What is most nearly the natural frequency of the control system?

(A)  0.707
(B)  3.12
(C)  5.92
(D)  9.75

Problems 36–42 assume that the natural frequency of the control system shown in Prob. 38 is 5 Hz.

**39.** What is most nearly the steady-state gain of the control system?

(A)  1.4
(B)  7.0
(C)  16
(D)  180

**40.** What is most nearly the damping ratio of the control system?

(A)  0.06
(B)  0.3
(C)  0.6
(D)  0.8

Problems 41 and 42 assume a damping ratio of 0.5.

**41.** What is most nearly the damped natural frequency of the control system?

(A)  0.866
(B)  3.54
(C)  3.75
(D)  4.33

**42.** What is most nearly the damped resonant frequency of the control system?

(A)  0.866
(B)  3.54
(C)  3.75
(D)  4.33

**43.** If the control system is given a unit step input, how will the response be characterized?

(A)  underdamped
(B)  overdamped
(C)  critically damped
(D)  unstable

**44.** A baud is

(A)  a measure of the intervals between transmitted digital data
(B)  a device used to demodulate audio data signals
(C)  a unit of magnitude of electronically transmitted audio signals
(D)  a unit of speed in digital data transmission measuring number of signals per second

**45.** The following code is an example of what programming technique?

```
b = cube(a)
cube(a)
{
b = a * a * a
return b
}
```

(A)  branching
(B)  function call
(C)  looping
(D)  subroutine

**46.** When can a professional provide services if conflicts of interest are involved?

(A)  if no compensation is received
(B)  if doing so does not personally benefit the professional
(C)  if full disclosure of potential conflicts are provided
(D)  never

**47.** When is the receipt of compensation from more than one party for the same project ethical?

(A) never
(B) when the total of all compensation received does not exceed "reasonable" compensation for the type of work performed
(C) always
(D) when all parties involved know about and agree to the arrangement

**48.** When may professionals make political donations?

(A) at no time
(B) if it is not for current, past, or future influence or favors
(C) if all donations are made as an individual and do not represent a firm or entity
(D) only below specified amounts

**49.** Which of the following is not always an ethics violation?

(A) signing plans or blueprints without having first designed and/or checked the plans
(B) revealing confidential information about a product without first obtaining permission
(C) granting a contract to a company for which the professional is an officer while concurrently serving on the board issuing the grants
(D) any individual accepting fees from contractors hired for a project

**50.** In the event of an ethical conflict, to whom does the professional hold the least ethical responsibility?

(A) the employer
(B) the client
(C) the consumer
(D) society

**51.** Ethical behavior is officially regulated by whom?

(A) individual employers
(B) registered engineers
(C) professional societies
(D) state enforcement agencies

**52.** Whistle blowing is

(A) an ethical practice
(B) an illegal practice
(C) an unethical practice
(D) a career-enhancing practice

**53.** Which of the following is not a reason professionals must adhere to a code of ethics?

(A) Professions are self-regulating.
(B) Professionals receive above-average compensation.
(C) Professionals have autonomy.
(D) Professions are responsible for training other professionals.

**54.** The National Society of Professional Engineers' (NSPE) Code of Ethics addresses competitive bidding. Which of the following is NOT stipulated?

(A) Engineers and their firms may refuse to bid competitively on engineering services.
(B) Clients are required to seek competitive bids for design services.
(C) Federal laws governing procedures for procuring engineering services (e.g., competitive bidding) remain in full force.
(D) Engineers and their societies may actively lobby for legislation that would prohibit competitive bidding for design services.

**55.** A credit card offers 1.2% effective monthly interest. What is most nearly the effective annual rate with monthly compounding?

(A) 7.9%
(B) 8.9%
(C) 14%
(D) 15%

**56.** A bank charges 12% simple interest on a $300 loan. Most nearly, how much will be repaid if the loan is paid back in one lump sum after three years?

(A) $108
(B) $408
(C) $415
(D) $421

**57.** Most nearly, how long will it take a sum of money to double at a 5% annual percentage rate?

(A) 6 years
(B) 10 years
(C) 11 years
(D) 14 years

**58.** A piece of machinery has an initial cost of $40,000 and results in an increase in annual maintenance costs of $2000. If the machinery saves the company $10,000 per year, in approximately how many years will the machine pay for itself if compounding is considered? The effective annual interest rate is 6%.

(A)  5.2 years
(B)  6.1 years
(C)  7 years
(D)  8 years

**59.** Funds are deposited in a savings account at an interest rate of 8% per annum. If the interest is compounded semi-annually, what is most nearly the initial amount that must be deposited to yield a total of $10,000 in 10 years?

(A)  $4530
(B)  $4560
(C)  $6730
(D)  $8200

**60.** $500 is deposited into a bank savings account with 6% interest compounded annually. Most nearly how much is in the account at the end of three years?

(A)  $550
(B)  $600
(C)  $650
(D)  $700

**61.** At the end of each year for five years, $500 is deposited into a credit union account. The credit union pays 5% interest compounded annually. At the end of five years (immediately following the fifth deposit) most nearly how much will be in the account?

(A)  $640
(B)  $1750
(C)  $2760
(D)  $3550

**62.** On January 1, $5000 is deposited into a high-interest savings account that pays 8% interest compounded annually. If all of the money is withdrawn in five equal end-of-year sums beginning December 31 of the first year, most nearly how much will each withdrawal be?

(A)  $1008
(B)  $1150
(C)  $1210
(D)  $1250

**63.** If you needed to have $800 in savings at the end of four years and your savings account yielded 5% interest paid annually, most nearly how much would you need to deposit today?

(A)  $570
(B)  $600
(C)  $660
(D)  $770

**64.** What is most nearly the reaction at A?

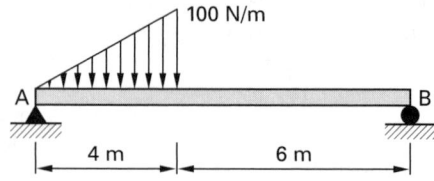

(A)  50 N
(B)  100 N
(C)  130 N
(D)  150 N

**65.** Find the magnitude of the force in the member marked with an "X." All members are pin-connected.

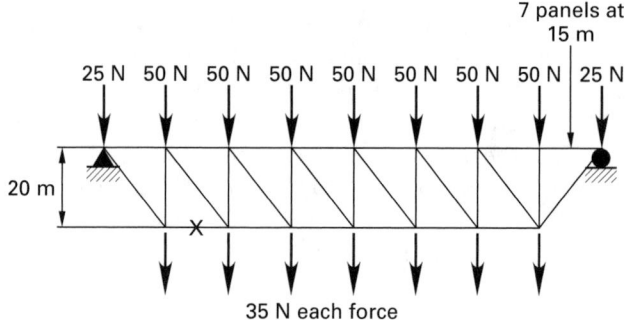

(A)  9.70 N
(B)  27.3 N
(C)  85.0 N
(D)  223 N

**66.** The approximate vertical force component in member BC is

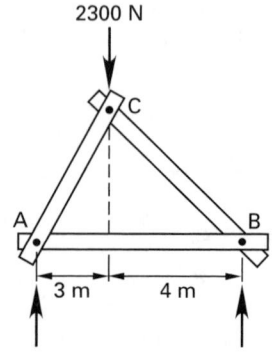

(A)  990 N
(B)  1300 N
(C)  2300 N
(D)  3600 N

**67.**  What is most nearly the frictional force between the 80 kg block and the ramp? The coefficient of static friction is 0.2, and the coefficient of dynamic friction is 0.15.

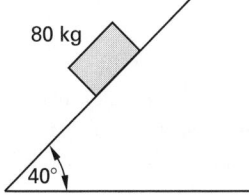

(A)  60 N
(B)  80 N
(C)  90 N
(D)  120 N

**68.**  Force $F$ is gradually increased until the 20 kg block begins moving to the right. The 5 kg block is prevented from moving by a cord. What is most nearly the minimum force $F$ for which movement is possible?

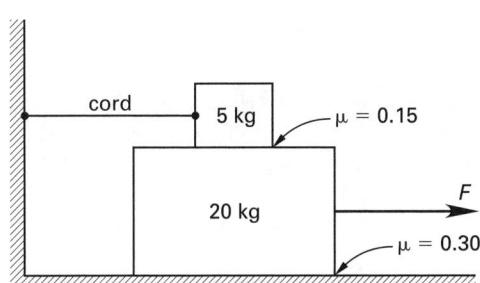

(A)  7.4 N
(B)  59 N
(C)  74 N
(D)  81 N

**69.**  Most nearly, what force, $F$, is required to lift a 50 N load? All pulleys are frictionless. Assume all strands are parallel.

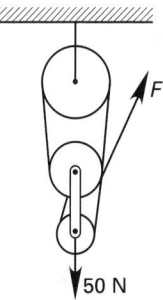

(A)  8.3 N
(B)  13 N
(C)  17 N
(D)  25 N

**70.**  What is most nearly the maximum value of $x$ such that $F$ can be applied without tipping the block? ($\mu = 0.4$.)

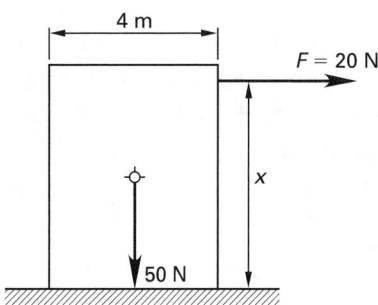

(A)  1.5 m
(B)  3.5 m
(C)  4.4 m
(D)  5.0 m

**71.**  What is most nearly the component of velocity perpendicular to the wall after impact if the coefficient of restitution is 0.8?

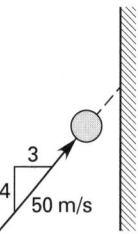

(A)  24 m/s
(B)  30 m/s
(C)  32 m/s
(D)  40 m/s

**72.**  The velocity of a particle at time $t$ is

$$v(t) = 12t^4 + \frac{7}{t}$$

Most nearly, what total distance is traveled between $t = 0.2$ and $t = 0.3$?

- (A) 0.98
- (B) 1.8
- (C) 2.8
- (D) 8.4

**73.** A spring has a spring constant of 10 N/cm. It is compressed 5 cm. The spring is released and pushes against a free projectile with a mass of 1 kg. The projectile velocity immediately after losing contact with the spring is most nearly

- (A) 0.32 m/s
- (B) 1.6 m/s
- (C) 32 m/s
- (D) 50 m/s

**74.** A rocket is moving through a vacuum. It changes its velocity from 9020 m/s to 5100 m/s in 48 s. The power required to accomplish this if the rocket's mass is 213 000 kg is most nearly

- (A) 34 GW
- (B) 120 GW
- (C) 170 GW
- (D) 250 GW

**75.** A projectile is launched at 52 degrees from horizontal with an initial velocity of 3600 m/s. If the mass of the projectile is 32 kg, what is most nearly the total kinetic and potential energy possessed by the projectile at $t = 13$ s? Neglect all forms of friction.

- (A) 5.9 kJ
- (B) 0.58 MJ
- (C) 210 MJ
- (D) 420 MJ

Problems 76 and 77 are based on the following statement.

A 30 cm long rod ($E = 3 \times 10^7$ N/cm$^2$, $\alpha = 6 \times 10^{-6}$ cm/cm·°C) with a 2 cm$^2$ cross section is fixed at both ends.

**76.** If the rod is heated to 60°C above the neutral temperature, what is most nearly the stress?

- (A) 110 N/cm$^2$
- (B) 11 000 N/cm$^2$
- (C) 36 000 N/cm$^2$
- (D) 57 000 N/cm$^2$

**77.** If one end of the rod is free to expand the elongation is most nearly

- (A) $5.4 \times 10^{-4}$ cm
- (B) $3.6 \times 10^{-4}$ cm
- (C) 0.01 cm
- (D) 0.03 cm

**78.** Vickers, Knoop, and Brinell are all names of

- (A) Nobel prize winners in metallurgy
- (B) thermodynamic constants
- (C) hardness tests
- (D) chi-squared statistics

**79.** What is the maximum flexural stress at a distance $x$ from the free end of a cantilever beam supporting a tip load, $P$?

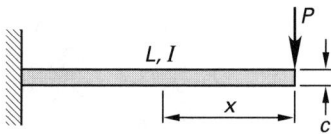

- (A) $\dfrac{Pxc}{2I}$
- (B) $\dfrac{Pc}{2I}$
- (C) $\dfrac{PcL}{2I}$
- (D) $\dfrac{Pxc}{2EI}$

**80.** What is most nearly the elongation in the cable if $F = 1000$ N? The cable's effective cross-sectional area is 2 cm$^2$. Its modulus of elasticity is $1.5 \times 10^6$ N/cm$^2$.

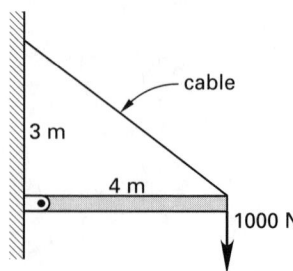

- (A) 0.0028 cm
- (B) 0.14 cm
- (C) 0.28 cm
- (D) 0.56 cm

**81.** A simply-supported beam carries a single concentrated load at its center. If its slenderness ratio is 150, it is most likely to fail

(A) where the moment is zero
(B) where the shear is maximum
(C) at a support
(D) where the slope of the deflection curve is zero

Problems 82 and 83 are based on the following statement and illustrations.

The results of a tensile test on a round specimen of a given material are shown.

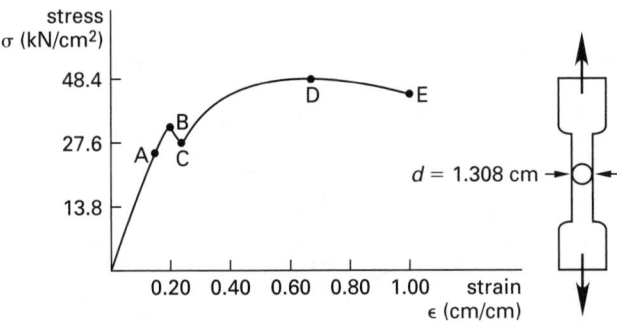

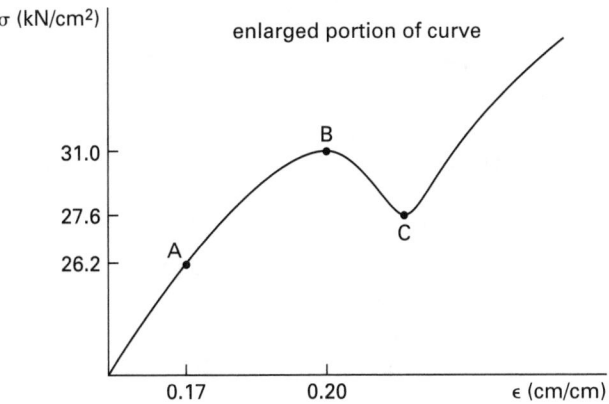

**82.** What is most nearly the yield stress?

(A) 26 kN/cm$^2$
(B) 28 kN/cm$^2$
(C) 29 kN/cm$^2$
(D) 31 kN/cm$^2$

**83.** What is most nearly the elastic limit of the material?

(A) 26 kN/cm$^2$
(B) 28 kN/cm$^2$
(C) 31 kN/cm$^2$
(D) 34 kN/cm$^2$

**84.** The aluminum rod shown in the following illustration has an initial diameter, $d_o$, of 30 mm and an initial gauge length, $\ell_o$, of 200 mm. The yield strength is 420 MPa. What is the modulus of elasticity if a force of 150 kN elongates the rod by 1.0 mm?

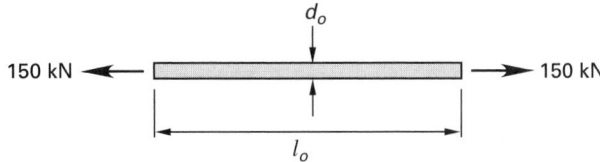

(A) 21 GPa
(B) 31 GPa
(C) 42 GPa
(D) 53 GPa

**85.** Which type of hardening will work to some extent in all metals?

(A) work-hardening
(B) annealing
(C) martempering
(D) austenitizing

**86.** What do impact tests determine?

(A) hardness
(B) yield strength
(C) toughness
(D) creep strength

**87.** What is one of the main differences between cast iron and steel?

(A) Steel has a lower carbon content.
(B) Steel always contains alloying metals such as nickel, chromium, manganese, and vanadium.
(C) Steel cannot be annealed, whereas cast iron can.
(D) Steel contains a large amount of uncombined carbon.

**88.** For a fixed curing time, the ultimate strength of concrete

(A) increases with a decrease in water content
(B) decreases with a decrease in water content
(C) is independent of water content if cured for a sufficiently long time
(D) is independent of curing pressure

**89.** The radius of a hypothetical electron orbit is known to be 0.75 Å. What is most nearly the de Broglie wavelength of the electron if four complete cycles constitute a stable pattern around the nucleus?

(A)  0.19 Å
(B)  1.2 Å
(C)  2.4 Å
(D)  4.7 Å

**90.** An orbital

(A)  may have 2, 8, 18, or 32 electrons
(B)  may have 2 electrons with the same spin direction
(C)  may be photographed with an electron microscope
(D)  may be unoccupied

**91.** For corrosion to occur, which of the following items must be present?

I.    anode
II.   cathode
III.  electrolyte

(A)  I and II
(B)  I and III
(C)  II and III
(D)  I, II, and III

**92.** A system consisting of a mixture of ice and water at a constant pressure is warmed from 0°C to 20°C. How many degrees of freedom does the system have?

(A)  −1
(B)  0
(C)  1
(D)  2

**93.** A venturi meter is used to measure air velocity. A one-fifth scale model of the venturi meter is built, and water is used as the test fluid. Viscosity of the air is $1.82 \times 10^{-5}$ N·s/m$^2$. Viscosity of the water is $9.82 \times 10^{-4}$ N·s/m$^2$. What will be the approximate ratio of the model to the actual velocities observed?

(A)  0.32
(B)  3.1
(C)  11
(D)  54

**94.** A 150 m long surface vessel is modeled at 1:50. Most nearly, what speed must the model travel if a 40 kph similarity is desired?

(A)  0.22 m/s
(B)  1.5 m/s
(C)  1.6 m/s
(D)  2.2 m/s

**95.** What is most nearly the hydraulic radius of an equilateral triangle (vertex down) open channel flowing at full capacity with a maximum depth of 3 m?

(A)  0.60 m
(B)  0.65 m
(C)  0.70 m
(D)  0.75 m

**96.** A 24 cm long rod floats vertically in water. It has a 1 cm$^2$ cross section and a specific gravity of 0.6. Most nearly, what length, $L$, is submerged?

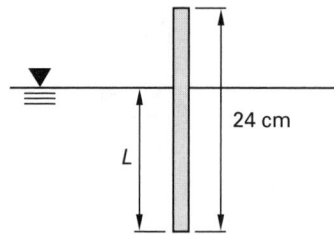

(A)  9.6 cm
(B)  14 cm
(C)  18 cm
(D)  19 cm

**97.** A tank is filled with water to a depth of 10 m. The total force on the gate is most nearly

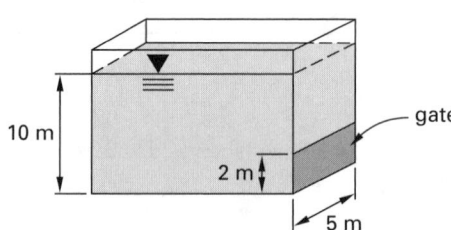

(A)  90 kN
(B)  440 kN
(C)  880 kN
(D)  980 kN

**98.** When a thin-bore, hollow glass tube is inserted into a container of mercury, the surface of the mercury in the tube

(A) is level with the surface of the mercury in the container
(B) is below the container surface due to cohesion
(C) is below the container surface due to adhesion
(D) is above the container surface due to cohesion

**99.** The fluid in a manometer tube is 60% water and 40% alcohol (specific gravity = 0.8). What is most nearly the manometer fluid height difference if a 42.7 kPa pressure difference is applied across the two ends of the manometer?

(A) 23 cm
(B) 47 cm
(C) 470 cm
(D) 550 cm

**100.** Carbon dioxide ($CO_2$) gas has a molecular weight of 44. The density of STP air is 1.29 kg/m$^3$. The specific gravity of $CO_2$ gas at 66°C and 138 kPa, using STP air as reference, is most nearly

(A) 0.67
(B) 1.1
(C) 1.7
(D) 2.1

**101.** A vacuum pump is used to drain a basement of 20°C water. The vapor pressure of water at this temperature is 2.34 kPa. The pump is incapable of lifting water higher than 10.5 m. The atmospheric pressure is most nearly

(A) 100 kPa
(B) 150 kPa
(C) 210 kPa
(D) 270 kPa

**102.** Six coulombs of charge pass through a wire in 2 s. What is most nearly the average current flowing?

(A) 1.6 A
(B) 3 A
(C) 4.8 A
(D) 6 A

**103.** An ideal transformer has 200 primary turns and 20 secondary turns. What is most nearly the secondary voltage if the primary voltage is 120 V?

(A) 1.2 V
(B) 12 V
(C) 120 V
(D) 1200 V

**104.** If 0.3 A flows in the secondary and 30 A flows in the primary of a perfectly matched, ideal transformer, what is most nearly the primary to secondary turns ratio?

(A) 1:100
(B) 1:10
(C) 10:1
(D) 100:1

**105.** Five watts are dissipated in a primary (input) circuit that includes a perfectly matched, ideal transformer with a primary-to-secondary turns ratio of 15:1. If the input resistance is 2000 Ω, what is most nearly the load resistance?

(A) 8.9 Ω
(B) 130 Ω
(C) 6.2 kΩ
(D) 30 kΩ

**106.** What is the total energy dissipated in the resistor when the capacitor discharges?

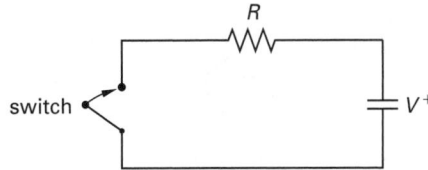

(A) $V^2 R$
(B) $\frac{1}{2}CV$
(C) $CV$
(D) $\frac{1}{2}CV^2$

**107.** What is most nearly the phase angle difference between the current and the voltage? Take the voltage as the reference.

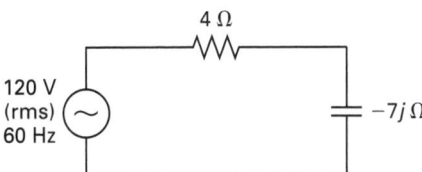

(A) −60°
(B) −35°
(C) −30°
(D) +30°

**108.** What is most nearly the phasor voltage drop across the current source?

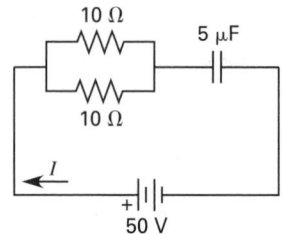

(A)  0.80∠16.8° V
(B)  1.2∠−53.7° V
(C)  5.0∠36.8° V
(D)  6.0∠−20.0° V

**109.**  Two 10 Ω resistances are connected in parallel. This combination is connected in series with a capacitor of 5 μF. The circuit is connected across a DC source voltage of 50 V. What is most nearly the maximum steady-state current through the battery?

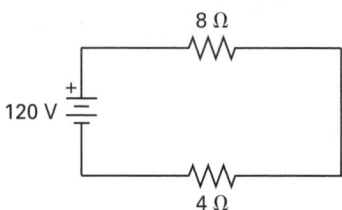

(A)  0
(B)  1 A
(C)  5 A
(D)  7 A

**110.**  The voltage appearing across the 4 Ω resistor is msot nearly

(A)  27 V
(B)  40 V
(C)  80 V
(D)  120 V

**111.**  A series $RLC$ circuit driven by an AC voltage contains reactances $X_L = 20$ Ω and $X_C = 14$ Ω and resistance $R = 10$ Ω. What is most nearly the impedance at resonant frequency?

(A)  10 Ω
(B)  14 Ω
(C)  20 Ω
(D)  24 Ω

**112.**  What are the changes in internal energy for reversible adiabatic and isothermal processes, respectively?

(A)  $C_p\Delta T$ and 0
(B)  0 and $C_v\Delta T$
(C)  $C_v\Delta T$ and $C_p\Delta T$
(D)  $C_v\Delta T$ and 0

**113.**  A steam engine operates between 150°C and 550°C. What is most nearly its theoretical maximum thermal efficiency?

(A)  27%
(B)  49%
(C)  73%
(D)  95%

**114.**  Regardless of the process, the change in enthalpy ($\Delta H$) for $n$ moles of an ideal gas is

(A)  $\dfrac{n\overline{R}T}{V}$
(B)  0
(C)  $nC_v\Delta T$
(D)  $nC_p\Delta T$

**115.**  Which of the following statements is true?

(A)  Entropy always decreases.
(B)  Entropy increases up to the critical temperature, and then it decreases.
(C)  Theoretically, entropy approaches zero as the temperature approaches zero.
(D)  None of the above statements are true.

**116.**  When is the equation $TdS = pdV + dU$ valid?

(A)  only in constant temperature processes
(B)  only in constant pressure processes
(C)  only in reversible processes
(D)  always

**117.**  Which of the following statements is true for a perfect gas flowing through an insulated valve?

(A)  Enthalpy is essentially unchanged.
(B)  Entropy decreases.
(C)  Temperature increases greatly.
(D)  Flow is isentropic.

**118.** If an initial volume of an ideal gas is compressed to one-half its original volume and to twice its original temperature, the pressure

(A) remains constant
(B) doubles
(C) quadruples
(D) halves

**119.** If an initial volume of saturated steam is expanded isothermally to twice the initial volume, the pressure

(A) decreases
(B) increases
(C) halves
(D) doubles

**120.** Which of the following is true of an adiabatic process?

(A) It allows heat transfer into the system but not out of the system.
(B) It may be reversible.
(C) It is one in which enthalpy remains unchanged.
(D) It is one in which the equation $W = Q$ is valid.

# STOP!

## DO NOT CONTINUE ON.

This concludes the Morning Section of the examination. If you finish before time is called, you may check your work on this section of the examination. You may not turn to the Afternoon Section of the exam until you are told to do so by your proctor. Be sure that all of your responses on the answer sheet are dark and completely fill the bubbles.

# Instructions

This section of the exam consists of 60 problems, each worth 2 points. You will have four hours in which to work this section. Your score will be determined by the number of problems that you solve correctly. No points will be deducted for incorrect answers. It is to your best advantage to try to answer every question.

When permission has been given by your proctor, break the seal on the Examination Booklet and remove the Answer Sheet. Write your name immediately in the space indicated. Check that all pages are present and legible. If any part of this Booklet is missing, your proctor will issue you a new Booklet.

All solutions must be entered on the Answer Sheet. No credit will be given for answers appearing only in the Examination Booklet. Mark your answers with the pencil provided. Marks must be dark and must completely fill the bubble. Record only one answer per problem; if you mark more than one answer, you will not receive credit for the problem. If you change an answer, be sure the old bubble is erased completely; incomplete erasures may be read as intended answers.

If you finish early, check your work and make sure you have correctly followed all instructions. After checking your answers, you may turn in your Examination Booklet and Answer Sheet and leave the examination room. Once you leave, you will not be permitted to return to work on your solutions.

WAIT FOR PERMISSION TO BEGIN.

Name: _____

       Last           First           Middle
                                       Initial

## FUNDAMENTALS OF ENGINEERING SAMPLE EXAMINATION

### EXAM 1
### AFTERNOON SECTION

#### Subject Breakdown

The major subject areas and their corresponding problem numbers are listed below.

| | |
|---|---|
| Advanced Engineering Mathematics | 1–6 |
| Engineering Probability and Statistics | 7–11 |
| Biology | 12–14 |
| Engineering Economics | 15–20 |
| Application of Engineering Mechanics | 21–28 |
| Engineering of Materials | 29–35 |
| Fluids | 36–44 |
| Electricity and Magnetism | 45–51 |
| Thermodynamics and Heat Transfer | 52–60 |

# Fundamentals of Engineering Sample Examination

## Afternoon Section

Name: _____

1.  (A) (B) (C) (D)     16. (A) (B) (C) (D)     31. (A) (B) (C) (D)     46. (A) (B) (C) (D)
2.  (A) (B) (C) (D)     17. (A) (B) (C) (D)     32. (A) (B) (C) (D)     47. (A) (B) (C) (D)
3.  (A) (B) (C) (D)     18. (A) (B) (C) (D)     33. (A) (B) (C) (D)     48. (A) (B) (C) (D)
4.  (A) (B) (C) (D)     19. (A) (B) (C) (D)     34. (A) (B) (C) (D)     49. (A) (B) (C) (D)
5.  (A) (B) (C) (D)     20. (A) (B) (C) (D)     35. (A) (B) (C) (D)     50. (A) (B) (C) (D)
6.  (A) (B) (C) (D)     21. (A) (B) (C) (D)     36. (A) (B) (C) (D)     51. (A) (B) (C) (D)
7.  (A) (B) (C) (D)     22. (A) (B) (C) (D)     37. (A) (B) (C) (D)     52. (A) (B) (C) (D)
8.  (A) (B) (C) (D)     23. (A) (B) (C) (D)     38. (A) (B) (C) (D)     53. (A) (B) (C) (D)
9.  (A) (B) (C) (D)     24. (A) (B) (C) (D)     39. (A) (B) (C) (D)     54. (A) (B) (C) (D)
10. (A) (B) (C) (D)     25. (A) (B) (C) (D)     40. (A) (B) (C) (D)     55. (A) (B) (C) (D)
11. (A) (B) (C) (D)     26. (A) (B) (C) (D)     41. (A) (B) (C) (D)     56. (A) (B) (C) (D)
12. (A) (B) (C) (D)     27. (A) (B) (C) (D)     42. (A) (B) (C) (D)     57. (A) (B) (C) (D)
13. (A) (B) (C) (D)     28. (A) (B) (C) (D)     43. (A) (B) (C) (D)     58. (A) (B) (C) (D)
14. (A) (B) (C) (D)     29. (A) (B) (C) (D)     44. (A) (B) (C) (D)     59. (A) (B) (C) (D)
15. (A) (B) (C) (D)     30. (A) (B) (C) (D)     45. (A) (B) (C) (D)     60. (A) (B) (C) (D)

Problems 1–3 are based on the following illustration.

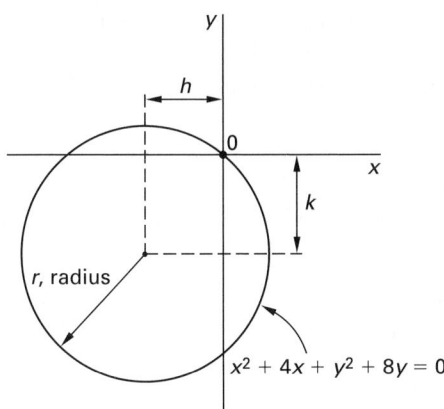

**1.** What are the coordinates of the circle's center?

(A) $(-4,-8)$
(B) $(2,-4)$
(C) $(-4,-2)$
(D) $(-2,-4)$

**2.** What is the radius of the circle?

(A) $\sqrt{8}$
(B) $2\sqrt{5}$
(C) $4\sqrt{5}$
(D) $10$

**3.** What is the slope of the line that is tangent to the circle and passes through the origin?

(A) $-2$
(B) $-3/4$
(C) $-1/2$
(D) $1/2$

Problems 4–6 are based on the following equation.

$$2y' = 3xy + 1$$

**4.** The equation is a

(A) second-order polynomial of two variables
(B) first-order, homogeneous equation
(C) linear first-order differential equation
(D) second-order differential equation

**5.** What is the integrating factor?

(A) $3x$
(B) $3/2x$
(C) $e^{-\frac{3}{2}x^2}$
(D) $e^{-\frac{3}{4}x^2}$

**6.** What is the solution?

(A) $y = \ln\left(\frac{3}{2}x^2\right) + C$
(B) $y = \frac{3}{2}x + C$
(C) $y = Ce^{\frac{3}{4}x^2} - \frac{1}{3}$
(D) $y = \left(e^{\frac{3}{4}x^2}\right)\left(\frac{1}{2}\int e^{-\frac{3}{4}x^2}\,dx + C\right)$

**7.** The least-squares method is used to plot a straight line through the data points $(3,-5)$, $(3,-2)$, $(4,3)$, and $(-1,6)$. The correlation coefficient is most nearly

(A) $-0.97$
(B) $-0.91$
(C) $-0.59$
(D) $-0.36$

**8.** Testing has shown that, on average, 2% of the bearings produced at a factory are defective. 20 bearings are chosen at random. The probability that exactly three of them are defective is most nearly

(A) 0.0037
(B) 0.0044
(C) 0.0059
(D) 0.0065

**9.** The number of different permutations of $n$ distinct objects taken $r$ at a time is

(A) $\dfrac{n!}{(n-r)!}$

(B) $\dfrac{n!}{r!(n-r)!}$

(C) $\dfrac{n!}{n_1!n_2!\ldots n_r!}$

(D) $\left(\dfrac{n+1}{2}\right)^r$

**10.** The equation $y = ax^3 + bx^2 + cx + d$ will fit how many constraints on a curve?

(A) one
(B) two
(C) three
(D) four

**11.** 100 random samples were taken from a large population. A particular numerical characteristic of sampled items was measured. The results of the measurements were as follows.

- 45 measurements were between 85.9 and 90.0

- 90.1 was observed once

- 90.2 was observed three times

- 90.3 was observed twice

- 90.4 was observed four times

- 45 measurements were between 90.5 and 95.8

The smallest value was 85.9 and the largest value was 95.8. The sum of all 100 measurements was 9117.0. Except those noted, no measurements occurred more than twice.

What is the mean of the measurements?

(A) 90.3
(B) 90.5
(C) 90.6
(D) 91.2

**12.** Protists are part of which biological group?

(A) seed plants
(B) eucaryotes
(C) eubacteria
(D) archaebacteria

**13.** What is the specific growth rate of an organismal growth batch culture that exhibits a constant change in cell concentration from $10^3$ mg/L to $10^5$ mg/L during the 15 h log growth phase?

(A) 0.07 h$^{-1}$
(B) 0.13 h$^{-1}$
(C) 0.3 h$^{-1}$
(D) 6.6 h$^{-1}$

**14.** A village with a stable population has a water supply that has been contaminated with benzene ($C_6H_6$) from a leaking underground storage tank. The leak occurred during the 20 yr prior to the tank being removed. The estimated average concentration of benzene during this period of leaking was 50 $\mu$g/L. The slope factor for benzene by the oral route is $2.0 \times 10^{-2}$ (mg/kg·day)$^{-1}$. Assume a 70 yr lifespan, 10 kg children, and 1 L/day child water consumption. What are the probable risks of additional cancers for the village's children who drank the water?

(A) children $0.71 \times 10^{-1}$
(B) children $1.4 \times 10^{-3}$
(C) children $2.9 \times 10^{-5}$
(D) children $5.7 \times 10^{-7}$

Problems 15–17 are based on the following statement.

A company must purchase a machine that will be used over the next 8 years. The purchase price is $10,000, and the salvage value after 8 years is $1000. The annual insurance is 2% of the purchase price, the electricity cost is $300 per year, and maintenance and replacement parts cost $100 per year. The effective annual interest rate for economic analysis is 6%. Neglect taxes.

**15.** What is most nearly the effective uniform annual cost of ownership?

(A) $1200
(B) $2100
(C) $2200
(D) $2300

**16.** What is most nearly the minimum acceptable average annual return on the machine?

(A) $2050
(B) $2110
(C) $2210
(D) $2330

**17.** What is the present worth of the machine if its return is $3000 per year for 8 years?

(A) $5530
(B) $8630
(C) $9260
(D) $24,000

**18.** A machine has an initial cost of $10,000 and an annual maintenance cost of $450. The life of the machine is 15 years, and its salvage value is $2500. Assuming an annual effective interest rate of 4%, the equivalent uniform annual cost of the machine is most nearly

(A) $1200
(B) $1400
(C) $1500
(D) $1800

**19.** A machine has an initial cost of $14,000, a life of 15 years, and a straight line depreciation value of $850. The salvage value is most nearly

(A) $850
(B) $1250
(C) $2400
(D) $3700

**20.** The salary range for a particular employee's job has six levels, each one 4% greater than the one below it. Circumstances dictate that the employee's salary must be reduced from the top (sixth) level to the second level. That results in a reduction of $140 per month. What is most nearly the employee's salary per month at level six?

(A) $840
(B) $960
(C) $980
(D) $3600

Problems 21–23 are based on the following statement and illustration. A simple truss mechanism is loaded as shown. All joints are pin-connected.

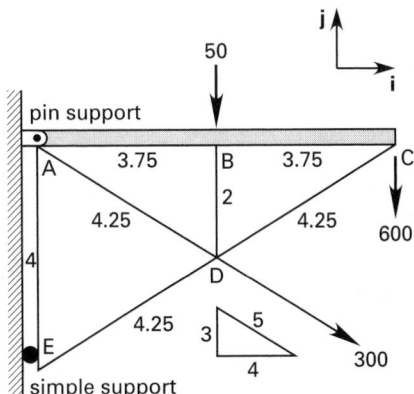

**21.** What is most nearly the reaction at pin A?

(A) $\mathbf{A} = 1461\mathbf{i} + 830\mathbf{j}$
(B) $\mathbf{A} = -1461\mathbf{i} + 830\mathbf{j}$
(C) $\mathbf{A} = -980\mathbf{i} - 730\mathbf{j}$
(D) $\mathbf{A} = +1400\mathbf{i} - 370\mathbf{j}$

**22.** What are most nearly the forces in members AD and ED?

(A) $F_{AD} = 676$ (tension)
$F_{ED} = 346$ (tension)
(B) $F_{AD} = 676$ (compression)
$F_{ED} = 346$ (compression)
(C) $F_{AD} = 380$ (tension)
$F_{ED} = 1383$ (compression)
(D) $F_{AD} = 380$ (compression)
$F_{ED} = 1383$ (tension)

**23.** Most nearly, what additional horizontal force applied at D will reduce the value of the reaction at roller E to zero?

(A) 1221 to the right
(B) 1461 to the right
(C) 2344 to the right
(D) 2441 to the right

**24.** The truss members are pin connected. If member BD is a round steel rod of radius 1.25 cm, what is most nearly the maximum load, $P$, that can be supported by the truss without causing the member to buckle? The rod's modulus of elasticity is 42.0 GPa.

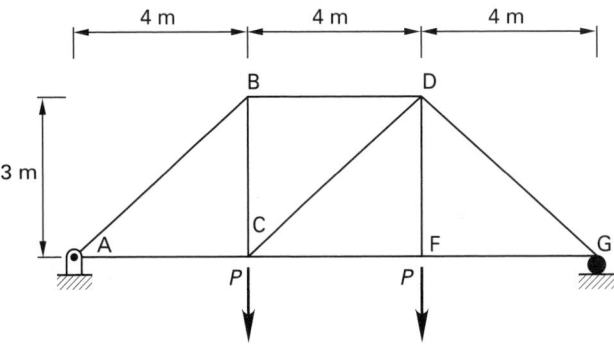

(A) 130 N
(B) 300 N
(C) 500 N
(D) 820 N

**25.** Determine the vertical deflection at the center of the beam. $EI$ is constant.

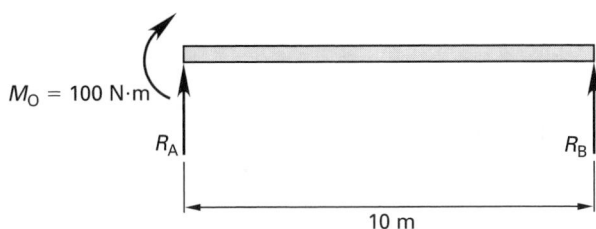

(A) $\dfrac{625}{EI}$ m downward

(B) $\dfrac{400}{EI}$ m downward

(C) $\dfrac{350}{EI}$ m downward

(D) $\dfrac{700}{EI}$ m upward

**26.** For an elastic, isotropic material, the modulus of elasticity, shear modulus, and Poisson's ratio are related by which of the following equations?

(A) $G = \dfrac{E}{2(1+v)}$

(B) $G = 2E(1+v)$

(C) $G = \dfrac{2E}{1+v}$

(D) $\dfrac{G}{2} = \dfrac{E}{1+v}$

**27.** A hollow, cylindrical copper column has an outer diameter of 5 cm and a wall thickness of 1 cm. A concentric load of 14.4 N is applied to one end. The axial stress is most nearly

(A) 0.76 Pa
(B) 19 Pa
(C) 76 Pa
(D) 190 Pa

**28.** Given a shear stress, $\tau_{xy}$, of 12 000 kPa and a shear modulus, $G$, of 87 GPa, the shear strain is most nearly

(A) $0.7 \times 10^{-5}$ rad
(B) $1.4 \times 10^{-4}$ rad
(C) $2.5 \times 10^{-4}$ rad
(D) $5.5 \times 10^{-4}$ rad

**29.** The hardness of steel may be increased by heating to approximately 1500°F and quenching in oil or water if

(A) the carbon content is from 0.2% to 2.0%
(B) the carbon content is above 3.0%
(C) the carbon content is below 0.2%
(D) all carbon is removed and the steel contains only chromium, nickel, manganese, or a combination of these

**30.** Plastically deforming a metal generally results in

(A) increased strength and ductility
(B) increased strength and decreased ductility
(C) decreased strength and ductility
(D) decreased strength and increased ductility

**31.** The following illustration depicts what type of metal after a tensile failure?

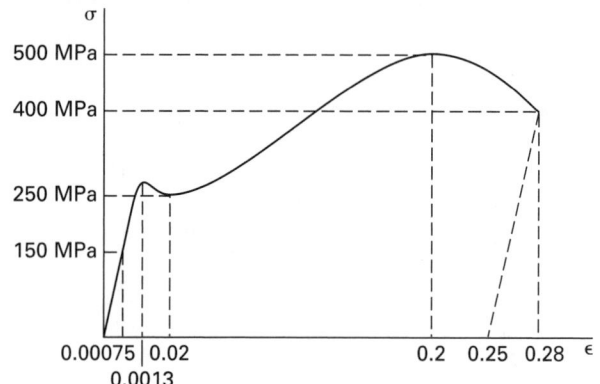

(A) very ductile metal
(B) moderately ductile metal
(C) brittle metal
(D) indeterminate

**32.** A ductile metal was submitted to a tensile test, resulting in the following stress-strain ($\sigma$-$\epsilon$) curve.

What is the percent elongation at failure?

(A) 2%
(B) 13%
(C) 25%
(D) 28%

**33.** The continuous yielding of a material under constant stress is known as

(A) failure
(B) strain
(C) creep
(D) ductility

**34.** The mass of an atom of nickel is most nearly

(A) $2.1 \times 10^{-23}$ g/atom
(B) $9.7 \times 10^{-23}$ g/atom
(C) $28 \times 10^{-23}$ g/atom
(D) $59 \times 10^{-23}$ g/atom

**35.** A 2 m long cylindrical bar of metal is suspended vertically from one end. The metal has a density of 2500 kg/m³ and modulus of elasticity of 100 GPa. The bar's cross-sectional area is 150 cm². What is the elongation of the bar due to its own mass?

(A)  $1.17 \times 10^{-10}$ m
(B)  $2.34 \times 10^{-9}$ m
(C)  $1.17 \times 10^{-8}$ m
(D)  $2.34 \times 10^{-7}$ m

**36.**  Most nearly, what total resultant force is acting on the inclined section of the water tank? Express your answer in kilonewtons per meter width of tank.

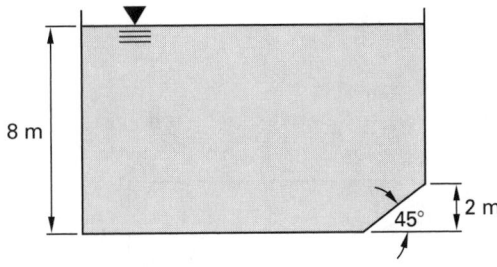

(A)  150 kN/m
(B)  190 kN/m
(C)  280 kN/m
(D)  310 kN/m

**37.**  What is most nearly the expected head loss per kilometer of closed circular pipe (17 cm inside diameter, friction factor of 0.03) when 3300 L/min of water flow under pressure?

(A)  0.053 m
(B)  22 m
(C)  53 m
(D)  4300 m

**38.**  A perfect venturi with a throat diameter of 1.8 cm is placed horizontally in a pipe with a 5 cm inside diameter. Eight kg of water flow through the pipe each second. What is most nearly the difference between the pipe and venturi throat static pressures?

(A)  30 kPa
(B)  490 kPa
(C)  640 kPa
(D)  970 kPa

**39.**  A static pressure gauge and mercury manometer are connected to a 50.8 cm pipeline flowing full of water. One cubic centimeter of mercury has a mass of 0.1336 N. What is most nearly the velocity at the center of the pipeline?

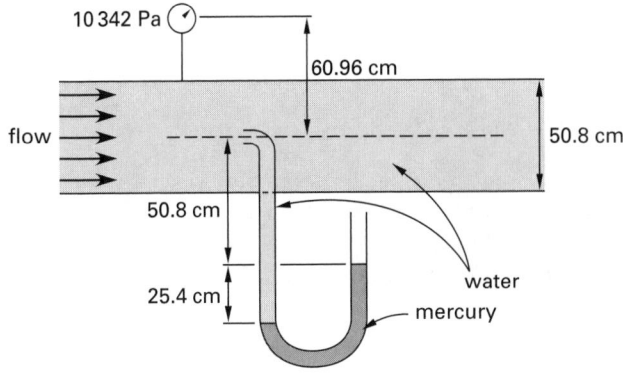

(A)  0.66 m/s
(B)  0.79 m/s
(C)  4.5 m/s
(D)  5.7 m/s

**40.**  Water is pumped up a hill into a reservoir by a pump at the bottom of the hill. The pump discharges water at a rate of 2 m/s and a pressure of 1000 kPa. Disregarding friction, the highest possible elevation of the reservoir's water surface is most nearly

(A)  100 m
(B)  1000 m
(C)  1020 m
(D)  10200 m

**41.**  Water flows in a 0.5 m (inside diameter) sewer line ($n = 0.015$, geometric slope $= 0.001$). The Manning coefficient, $n$, varies with depth. Flow is full, uniform, and steady. The flow rate is most nearly

(A)  0.10 m$^3$/s
(B)  0.21 m$^3$/s
(C)  0.53 m$^3$/s
(D)  0.70 m$^3$/s

**42.**  The velocities upstream and downstream, $v_1$ and $v_2$ respectively, of a 4.0 m wide sluice gate are unknown. The upstream depth is 2.0 m, and the downstream depth is 0.6 m. Flow is uniform and steady. The geometric slope across the sluice gate is zero. The downstream velocity, $v_2$, is most nearly

(A)  0.3 m/s
(B)  3 m/s
(C)  4.7 m/s
(D)  5.5 m/s

**43.**  A rectangular channel ($n = 0.013$, $s = 0.004$) has a depth of 3 m. The width of the channel is 5 m. The velocity of water in the channel is most nearly

(A) 1 m/s
(B) 6 m/s
(C) 15 m/s
(D) 90 m/s

**44.** A rectangular channel on a 0.002 slope is constructed of finished concrete ($n = 0.012$). The channel is 2.4 m wide with a depth of 1.5 m. The flow rate is most nearly

(A) $0.67 \text{ m}^3/\text{s}$
(B) $2.9 \text{ m}^3/\text{s}$
(C) $10 \text{ m}^3/\text{s}$
(D) $41 \text{ m}^3/\text{s}$

Problems 45–48 are based on the following information and illustration.

single-phase transformer
90 kVA
10,000 V primary: 2000 V secondary
60 Hz
power factor: 0.8 leading

$$R_1 = 6 \ \Omega \qquad R_2' = 7 \ \Omega$$
$$X_{L_1} = 31 \ \Omega \qquad X_{L_2}' = 31 \ \Omega$$
$$X_m' = 55 \text{ k}\Omega \qquad R_c' = 120 \text{ k}\Omega$$
$$|Z_L| = 55 \ \Omega$$

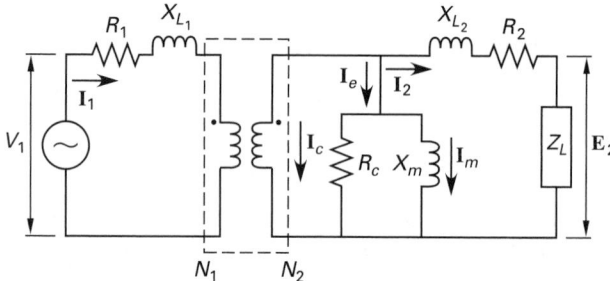

**45.** What is $\mathbf{I}_2$ referred to the secondary side if the secondary voltage is 2000 V?

(A) $36.36\angle-36.87°$ A
(B) $36.87\angle-36.36°$ A
(C) $33.10\angle36.87°$ A
(D) $36.36\angle36.87°$ A

**46.** What is $\mathbf{E}_2$ referred to the secondary side?

(A) $2.0\angle0°$ kV
(B) $5.0\angle0°$ kV
(C) $2.0\angle90°$ kV
(D) $110\angle90°$ kV

**47.** What is $\mathbf{I}_2$ referred to the primary side?

(A) $36.97\angle-36.36°$ A
(B) $7.27\angle-36.87°$ A
(C) $7.27\angle36.87°$ A
(D) $181\angle36.87°$ A

**48.** What is $\mathbf{E}_2$ referred to the primary side?

(A) $0.80\angle0°$ kV
(B) $5.0\angle0°$ kV
(C) $10\angle0°$ kV
(D) $50\angle0°$ kV

Problems 49–51 are based on the following illustration.

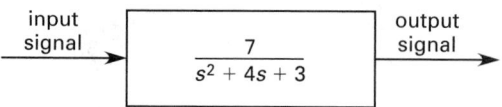

**49.** What is the output signal as a function of time if a unit impulse function is the input signal?

(A) $\left(\dfrac{7}{2}\right)\left(e^t + e^{-3t}\right)$

(B) $\left(\dfrac{7}{2}\right)\left(e^{-t} + e^{-3t}\right)$

(C) $\left(\dfrac{7}{2}\right)\left(e^{-3t} - e^{-t}\right)$

(D) $\left(\dfrac{7}{2}\right)\left(e^{-t} - e^{-3t}\right)$

**50.** What is the steady-state output if the unit step function of height 5 at $t = 0$ is the input signal?

(A) 3/35
(B) 35/6
(C) 35/3
(D) 7/3

**51.** What is the steady-state response if the sinusoid $7\sin(2t + \pi/3)$ is the input signal?

(A) $6.08\angle-157°$
(B) $6.08\angle-37°$
(C) $6.08\angle52.9°$
(D) $6.08\angle142.9°$

**52.** What is most nearly the coefficient of performance of a Carnot refrigeration cycle operating between $-23.3°$C and $-123.3°$C?

(A) −0.50
(B) 0.50
(C) 1.1
(D) 1.5

**53.** Most nearly, what work is done in compressing 1 kg of air originally at standard temperature and pressure (0°C, 1 atm) to one-half of its original volume in an isothermal process?

(A) 38 kJ
(B) 53 kJ
(C) 76 kJ
(D) 150 kJ

Problems 54–59 are based on the following statement and illustration.

A 1 MW power plant operates on the simple turbine cycle illustrated above. The isentropic efficiencies of the pump and turbine are 60% and 85.7%, respectively.

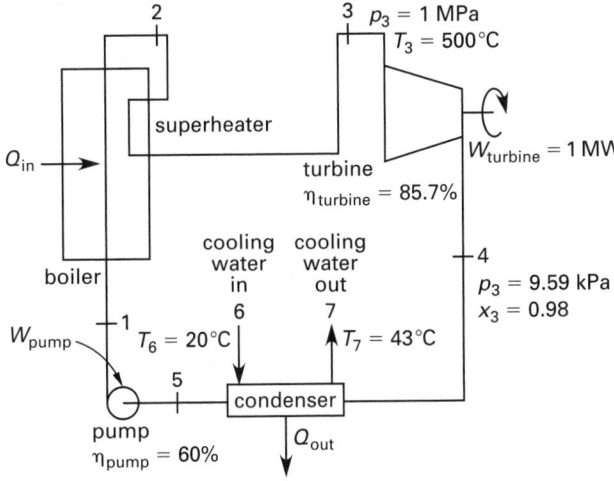

**54.** Superheated steam enters the turbine at 1 MPa and 500°C. Steam leaves the turbine at 9.59 kPa and with a quality of 98%. The ideal turbine work per unit mass of steam is most nearly

(A) 650 kJ/kg
(B) 810 kJ/kg
(C) 940 kJ/kg
(D) 1100 kJ/kg

**55.** Most nearly, what mass of steam passing through the turbine each hour would be required to generate 1 MW of power?

(A) 1900 kg/h
(B) 3100 kg/h
(C) 3800 kg/h
(D) 4500 kg/h

**56.** The enthalpy of the liquid leaving the condenser is most nearly

(A) 170 kJ/kg
(B) 180 kJ/kg
(C) 190 kJ/kg
(D) 200 kJ/kg

**57.** Cooling water enters the condenser at 20°C and leaves at 43°C. What is most nearly the flow rate of the cooling water?

(A) $8.7 \times 10^3$ kg/h
(B) $2.9 \times 10^4$ kg/h
(C) $1.1 \times 10^5$ kg/h
(D) $1.2 \times 10^5$ kg/h

**58.** The energy per unit mass of condensate added by the boiler feed pump is most nearly

(A) 800 J/kg
(B) 1000 J/kg
(C) 1700 J/kg
(D) 2400 J/kg

**59.** What is the approximate temperature of the water leaving the boiler feed pump?

(A) 45.2°C
(B) 45.4°C
(C) 46.1°C
(D) 47.0°C

**60.** In an isentropic process, $p_1 = 7.25$ N/cm$^2$, $p_2 = 10.73$ N/cm$^2$, and $T_1 = 355$K. The ratio of specific heats is 1.4. What is most nearly the value of $T_2$?

(A) 330K
(B) 360K
(C) 400K
(D) 440K

# Instructions

This section of the exam consists of 120 problems, each worth 1 point. You will have four hours in which to work this section. Your score will be determined by the number of problems that you solve correctly. No points will be deducted for incorrect answers. It is to your best advantage to try to answer every question.

When permission has been given by your proctor, break the seal on the Examination Booklet and remove the Answer Sheet. Write your name immediately in the space indicated. Check that all pages are present and legible. If any part of this Booklet is missing, your proctor will issue you a new Booklet.

All solutions must be entered on the Answer Sheet. No credit will be given for answers appearing only in the Examination Booklet. Mark your answers with the pencil provided. Marks must be dark and must completely fill the bubble. Record only one answer per problem; if you mark more than one answer, you will not receive credit for the problem. If you change an answer, be sure the old bubble is erased completely; incomplete erasures may be read as intended answers.

If you finish early, check your work and make sure you have correctly followed all instructions. After checking your answers, you may turn in your Examination Booklet and Answer Sheet and leave the examination room. Once you leave, you will not be permitted to return to work on your solutions.

Do not work any problems from the Afternoon Section of the exam during the first four hours of this exam.

WAIT FOR PERMISSION TO BEGIN.

Name: _____

       Last      First      Middle
                             Initial

## FUNDAMENTALS OF ENGINEERING SAMPLE EXAMINATION

### EXAM 2
### MORNING SECTION

#### Subject Breakdown

The major subject areas and their corresponding problem numbers are listed below.

| Subject | Problems |
|---|---|
| Mathematics | 1–18 |
| Engineering and Probability Statistics | 19–26 |
| Chemistry | 27–36 |
| Computers | 37–45 |
| Ethics and Business Practices | 46–54 |
| Engineering Economics | 55–63 |
| Engineering Mechanics (Statics and Dynamics) | 64–75 |
| Strength of Materials | 76–84 |
| Material Properties | 85–92 |
| Fluid Mechanics | 93–101 |
| Electricity and Magnetism | 102–111 |
| Thermodynamics | 112–120 |

# Fundamentals of Engineering Sample Examination

## Morning Section

Name: _____

| | | | | | | | | | | | |
|---|---|---|---|---|---|---|---|---|---|---|---|
| 1. | (A) (B) (C) (D) | 31. | (A) (B) (C) (D) | 61. | (A) (B) (C) (D) | 91. | (A) (B) (C) (D) |
| 2. | (A) (B) (C) (D) | 32. | (A) (B) (C) (D) | 62. | (A) (B) (C) (D) | 92. | (A) (B) (C) (D) |
| 3. | (A) (B) (C) (D) | 33. | (A) (B) (C) (D) | 63. | (A) (B) (C) (D) | 93. | (A) (B) (C) (D) |
| 4. | (A) (B) (C) (D) | 34. | (A) (B) (C) (D) | 64. | (A) (B) (C) (D) | 94. | (A) (B) (C) (D) |
| 5. | (A) (B) (C) (D) | 35. | (A) (B) (C) (D) | 65. | (A) (B) (C) (D) | 95. | (A) (B) (C) (D) |
| 6. | (A) (B) (C) (D) | 36. | (A) (B) (C) (D) | 66. | (A) (B) (C) (D) | 96. | (A) (B) (C) (D) |
| 7. | (A) (B) (C) (D) | 37. | (A) (B) (C) (D) | 67. | (A) (B) (C) (D) | 97. | (A) (B) (C) (D) |
| 8. | (A) (B) (C) (D) | 38. | (A) (B) (C) (D) | 68. | (A) (B) (C) (D) | 98. | (A) (B) (C) (D) |
| 9. | (A) (B) (C) (D) | 39. | (A) (B) (C) (D) | 69. | (A) (B) (C) (D) | 99. | (A) (B) (C) (D) |
| 10. | (A) (B) (C) (D) | 40. | (A) (B) (C) (D) | 70. | (A) (B) (C) (D) | 100. | (A) (B) (C) (D) |
| 11. | (A) (B) (C) (D) | 41. | (A) (B) (C) (D) | 71. | (A) (B) (C) (D) | 101. | (A) (B) (C) (D) |
| 12. | (A) (B) (C) (D) | 42. | (A) (B) (C) (D) | 72. | (A) (B) (C) (D) | 102. | (A) (B) (C) (D) |
| 13. | (A) (B) (C) (D) | 43. | (A) (B) (C) (D) | 73. | (A) (B) (C) (D) | 103. | (A) (B) (C) (D) |
| 14. | (A) (B) (C) (D) | 44. | (A) (B) (C) (D) | 74. | (A) (B) (C) (D) | 104. | (A) (B) (C) (D) |
| 15. | (A) (B) (C) (D) | 45. | (A) (B) (C) (D) | 75. | (A) (B) (C) (D) | 105. | (A) (B) (C) (D) |
| 16. | (A) (B) (C) (D) | 46. | (A) (B) (C) (D) | 76. | (A) (B) (C) (D) | 106. | (A) (B) (C) (D) |
| 17. | (A) (B) (C) (D) | 47. | (A) (B) (C) (D) | 77. | (A) (B) (C) (D) | 107. | (A) (B) (C) (D) |
| 18. | (A) (B) (C) (D) | 48. | (A) (B) (C) (D) | 78. | (A) (B) (C) (D) | 108. | (A) (B) (C) (D) |
| 19. | (A) (B) (C) (D) | 49. | (A) (B) (C) (D) | 79. | (A) (B) (C) (D) | 109. | (A) (B) (C) (D) |
| 20. | (A) (B) (C) (D) | 50. | (A) (B) (C) (D) | 80. | (A) (B) (C) (D) | 110. | (A) (B) (C) (D) |
| 21. | (A) (B) (C) (D) | 51. | (A) (B) (C) (D) | 81. | (A) (B) (C) (D) | 111. | (A) (B) (C) (D) |
| 22. | (A) (B) (C) (D) | 52. | (A) (B) (C) (D) | 82. | (A) (B) (C) (D) | 112. | (A) (B) (C) (D) |
| 23. | (A) (B) (C) (D) | 53. | (A) (B) (C) (D) | 83. | (A) (B) (C) (D) | 113. | (A) (B) (C) (D) |
| 24. | (A) (B) (C) (D) | 54. | (A) (B) (C) (D) | 84. | (A) (B) (C) (D) | 114. | (A) (B) (C) (D) |
| 25. | (A) (B) (C) (D) | 55. | (A) (B) (C) (D) | 85. | (A) (B) (C) (D) | 115. | (A) (B) (C) (D) |
| 26. | (A) (B) (C) (D) | 56. | (A) (B) (C) (D) | 86. | (A) (B) (C) (D) | 116. | (A) (B) (C) (D) |
| 27. | (A) (B) (C) (D) | 57. | (A) (B) (C) (D) | 87. | (A) (B) (C) (D) | 117. | (A) (B) (C) (D) |
| 28. | (A) (B) (C) (D) | 58. | (A) (B) (C) (D) | 88. | (A) (B) (C) (D) | 118. | (A) (B) (C) (D) |
| 29. | (A) (B) (C) (D) | 59. | (A) (B) (C) (D) | 89. | (A) (B) (C) (D) | 119. | (A) (B) (C) (D) |
| 30. | (A) (B) (C) (D) | 60. | (A) (B) (C) (D) | 90. | (A) (B) (C) (D) | 120. | (A) (B) (C) (D) |

**1.** What is the slope of the line tangent to the parabola $y = 12x^2 + 3$ at a point where $x = 5$?

(A) 24
(B) 120
(C) 303
(D) 515

**2.** Which of the following equations correctly describes the shaded area of the $x$-$y$ plane?

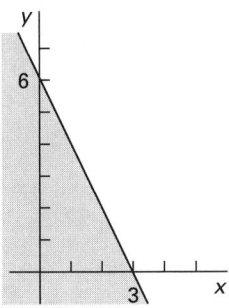

(A) $2x - y \leq 6$
(B) $2x + y \leq 6$
(C) $2x - y \geq 6$
(D) $x + 2y \geq 6$

**3.** What is the $(x, y)$ solution for the following system of two simultaneous equations?

$$3x - 6y = 7$$
$$2x - 11y = -5$$

(A) $\left( \dfrac{107}{21}, \dfrac{29}{21} \right)$

(B) $\left( -\dfrac{17}{45}, \dfrac{1}{45} \right)$

(C) $\left( -\dfrac{107}{21}, \dfrac{29}{21} \right)$

(D) $\left( \dfrac{107}{45}, -\dfrac{106}{45} \right)$

**4.** What is the simplified equivalent expression of

$$(\cot^2 \theta)(\sin^2 \theta) + \frac{1}{\csc^2 \theta}$$

(A) $2 \sin^2 \theta$
(B) $2 \cos \theta$
(C) $2 \cos^2 \theta$
(D) $1$

**5.** What is the derivative with respect to $x$ of $\sqrt{2x + 9x^2}$?

(A) $\dfrac{1 + 9x}{\sqrt{2x + 9x^2}}$

(B) $9x + 1$

(C) $\sqrt{2 + 18x}$

(D) $(9x + 1)\sqrt{2x + 9x^2}$

**6.** What is the limit?

$$\lim_{x \to \infty} \left( \frac{10x^2 - 5x + 1}{(5x - 3)(2x)} \right)$$

(A) $-1/3$
(B) $0$
(C) $5/6$
(D) $1$

**7.** What is the integral $\int x(x + 1)dx$?

(A) $\dfrac{x^3}{3} + x + C$

(B) $\dfrac{x^3}{3} + \dfrac{x^2}{2}$

(C) $\dfrac{x^3}{3} + \dfrac{x^2}{2} + C$

(D) $x^3 + x^2 + C$

**8.** How many significant digits are there in the number 023059.11005?

(A) 5
(B) 9
(C) 10
(D) 11

**9.** The horizontal angle from the ground to the top of a palm tree some unknown distance away is 46.18°. At a point 40 m directly behind the first point, the horizontal angle to the top of the tree is 29.23°. What is most nearly the distance from the palm tree to the first point?

(A) 42 m
(B) 46 m
(C) 51 m
(D) 61 m

**10.** What is most nearly the acute angle between vectors $\mathbf{V}_1 = (3, 2, 1)$ and $\mathbf{V}_2 = (2, 3, 2)$ based at the origin?

(A)  25°
(B)  33°
(C)  35°
(D)  59°

**11.**  What is most nearly the interior angle, $\theta$, of a regular polygon with seven sides?

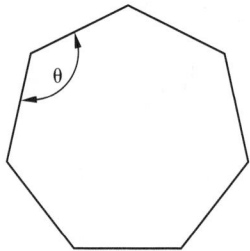

(A)  51°
(B)  64°
(C)  116°
(D)  129°

**12.**  Which of the following statements is true concerning the Taylor series expansion for $\cos x$?

(A)  The series contains only odd powers of $x$.
(B)  The series contains only even powers of $x$.
(C)  The series contains only negative odd powers of $x$.
(D)  The series contains only every other odd power of $x$.

**13.**  $A$ and $B$ are subsets of $Q$.

$$A = (4, 7, 9)$$
$$B = (4, 5, 9, 10)$$
$$Q = (4, 5, 6, 7, 8, 9, 10)$$

What is $\overline{A} \cup B$?

(A)  (4, 5, 6, 7, 8, 9, 10)
(B)  (4, 5, 7, 9, 10)
(C)  (4, 5, 6, 8, 9, 10)
(D)  (5, 10)

**14.**  What is the total area inside the cardioid, $r$?

$$r = a(1 + \cos\theta)$$

(A)  $\dfrac{3\pi a^2}{2}$

(B)  $\dfrac{2\pi a^2}{3}$

(C)  $\dfrac{3\pi a^2}{4}$

(D)  $\dfrac{4\pi a^2}{3}$

**15.**  What are the values of $B_1$ and $B_2$?

$$\begin{bmatrix} 9 & 7 \\ 1 & 3 \end{bmatrix} \begin{bmatrix} B_1 \\ B_2 \end{bmatrix} = \begin{bmatrix} 2 \\ 1 \end{bmatrix}$$

(A)  $B_1 = -\dfrac{1}{20}; \ B_2 = \dfrac{7}{20}$

(B)  $B_1 = \dfrac{7}{20}; \ B_2 = -\dfrac{1}{20}$

(C)  $B_1 = -\dfrac{1}{20}; \ B_2 = -\dfrac{7}{20}$

(D)  $B_1 = 10; \ B_2 = 20$

Problems 16 and 17 are based on the following equation.

$$xy' + 3x - 1 = 0$$

**16.**  What type of differential equation is this?

(A)  homogeneous
(B)  second-order linear
(C)  non-linear
(D)  first-order separable

**17.**  The solution for this equation is

(A)  $y = \ln(1 - 3x) + C$

(B)  $y = \dfrac{1}{x} - 3 + C$

(C)  $y = \ln x - 3x + C$

(D)  $y = \ln x - \dfrac{3x^2}{2} + C$

**18.**  Consider three vectors, $\overline{A}$, $\overline{B}$, and $\overline{C}$, with the following two properties.

$$\overline{A} \cdot (\overline{B} \times \overline{C}) = 0$$
$$\overline{B} \perp \overline{C}$$

Which of the following must be true?

(A) $\overline{A} \parallel \overline{B}$
(B) $\overline{A}$, $\overline{B}$, and $\overline{C}$ are coplanar
(C) $\overline{A} \parallel \overline{C}$
(D) $\overline{A} \perp$ to $\overline{B}$ or $\overline{C}$

**19.** Seven measurements are taken: 4.31, 4.39, 4.38, 4.33, 4.36, 4.32, and 4.37. What is most nearly the sample standard deviation?

(A) 0.0155
(B) 0.0167
(C) 0.0291
(D) 0.0313

**20.** Six design engineers are eligible for promotion to pay grade G8, but only four spots are available. How many different combinations of promoted engineers are possible?

(A) 4
(B) 6
(C) 15
(D) 20

**21.** A coin is tossed three times. What is the approximate probability of heads appearing at least one time?

(A) 0.67
(B) 0.75
(C) 0.80
(D) 0.88

**22.** Six board members of an engineering firm meet in the conference room. The conference table is round and has six identical seats. How many unique seating combinations exist? (Do not count rotations. Shifting each person one seat to the left would not result in a new combination.)

(A) 24
(B) 120
(C) 240
(D) 720

**23.** A deck of ten children's cards contains three fish cards, two dog cards, and five cat cards. What is the probability of drawing either a cat card or a dog card from a full deck?

(A) 1 in 10
(B) 2 in 10
(C) 5 in 10
(D) 7 in 10

**24.** What is approximately the sample standard deviation of the data set {17, 18, 24, and 33}?

(A) 5.8
(B) 6.5
(C) 7.3
(D) 8.0

**25.** On average, a furniture store sells four card tables in a week. Assuming a Poisson distribution for the weekly sales, the probability that the store will sell exactly seven card tables in a given week is most nearly

(A) 0.060
(B) 0.075
(C) 0.11
(D) 0.15

**26.** The least squares method is used to plot a straight line through the data points (2,10), (4,9), (6,6), and (7,4). The slope of the line is most nearly

(A) 1.1
(B) 1.2
(C) 1.5
(D) 1.7

Problems 27–31 are based on the following formulas.

I. $C_2H_2$
II. $(CH_3)_2O$
III. $C_2H_5Br$
IV. $C_2H_5COOH$
V. $C_2H_4$

**27.** Which formula(s) describe a carboxylic acid?

(A) III and IV
(B) IV only
(C) II and IV
(D) III only

**28.** Which formula(s) describe an alkane?

(A) III only
(B) V only
(C) I and V
(D) none of the above

**29.** Which formula(s) describe an alkene?

(A) I only
(B) V only
(C) I and V
(D) III only

**30.** Which formula(s) describe an ether?

(A) IV only
(B) II only
(C) II and IV
(D) III only

**31.** Which formula(s) describe an alkyl halide?

(A) II only
(B) III only
(C) II and III
(D) I only

**32.** A hydrate is a

(A) buffer solution of water
(B) compound in which hydrogen is combined with an element less electronegative than itself
(C) compound containing a definite number of water molecules in its chemical composition
(D) salt containing the hydroxyl radical

**33.** Two moles of hydrogen react with 1 mole of oxygen to produce which of the following?

(A) 1 mole of water
(B) 2 moles of water
(C) 3 moles of water
(D) 2 moles of water with 1 mole of hydrogen left over

**34.** Ten kilograms of hydrogen gas ($H_2$) are mixed with 355 kg of chlorine ($Cl_2$) in a 0.50 m$^3$ drum. The two gases react to produce hydrogen chloride. What is the final pressure in the drum if the final temperature is 60°C?

(A) 0.055 MPa
(B) 2.0 MPa
(C) 10 MPa
(D) 55 MPa

Problems 35 and 36 are based on the following statement.

Ten kilograms of oxygen and 2 kg of hydrogen are mixed in a 1 m$^3$ vessel. The mixture is at 300K, and initially no reaction takes place.

**35.** What is most nearly the partial pressure of the hydrogen?

(A) 2.5 MPa
(B) 3.4 MPa
(C) 5.0 MPa
(D) 12 MPa

**36.** A spark causes the hydrogen and oxygen to react to form water. Assuming complete combustion, approximately how much water is produced?

(A) 5.6 kg
(B) 11 kg
(C) 12 kg
(D) 18 kg

**37.** What is most nearly the overall gain of the cascaded system?

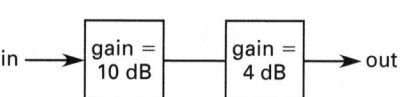

(A) 1.6 dB
(B) 12 dB
(C) 14 dB
(D) 40 dB

**38.** What is most nearly the overall gain, $x_o/x_i$, of the following positive feedback system?

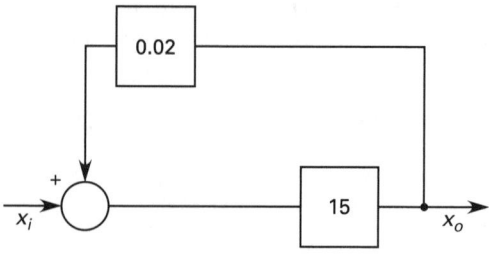

(A) 12
(B) 15
(C) 18
(D) 21

**39.** Which of the following systems is controllable?

(A) $\begin{bmatrix} x_1(t) \\ x_2(t) \end{bmatrix} = \begin{bmatrix} 1 & 4 \\ 2 & 8 \end{bmatrix} \begin{bmatrix} x_1(t) \\ x_2(t) \end{bmatrix} + \begin{bmatrix} 1 \\ 2 \end{bmatrix} \begin{bmatrix} u_1(t) \\ u_2(t) \end{bmatrix}$

(B) $\begin{bmatrix} x_1(t) \\ x_2(t) \end{bmatrix} = \begin{bmatrix} 1 & -5 \\ -2 & 10 \end{bmatrix} \begin{bmatrix} x_1(t) \\ x_2(t) \end{bmatrix} + \begin{bmatrix} 10 \\ 2 \end{bmatrix} \begin{bmatrix} u_1(t) \\ u_2(t) \end{bmatrix}$

(C) $\begin{bmatrix} x_1(t) \\ x_2(t) \end{bmatrix} = \begin{bmatrix} 1 & 3 \\ -2 & 6 \end{bmatrix} \begin{bmatrix} x_1(t) \\ x_2(t) \end{bmatrix} + \begin{bmatrix} 10 \\ 10 \end{bmatrix} \begin{bmatrix} u_1(t) \\ u_2(t) \end{bmatrix}$

(D) $\begin{bmatrix} x_1(t) \\ x_2(t) \end{bmatrix} = \begin{bmatrix} 0 & 3 \\ 0 & 4 \end{bmatrix} \begin{bmatrix} x_1(t) \\ x_2(t) \end{bmatrix} + \begin{bmatrix} 0 \\ 4 \end{bmatrix} \begin{bmatrix} u_1(t) \\ u_2(t) \end{bmatrix}$

Problems 40–42 are based on the following statement and illustration.

The system whose block diagram is shown is governed by the following differential equation.

$$C(t) = 10r(t) - 16C(t)$$
$$C(0) = 0$$

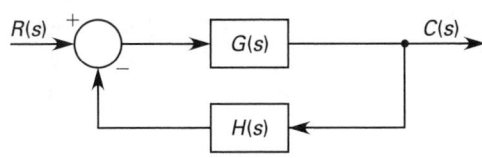

**40.** What is the transfer function, $G(s) = C(s)/R(s)$, for the system?

(A) $\dfrac{16}{s - 10}$

(B) $\dfrac{10}{s + 16}$

(C) $10s + 16$

(D) $\dfrac{s + 16}{10}$

**41.** Which of the following is an expression for $H(s)$?

(A) $s + 15$

(B) $\dfrac{s}{10} + \dfrac{3}{2}$

(C) $8/5$

(D) $8s$

**42.** What value does $C(t)$ approach as $t$ approaches infinity if $r(t)$ is given by $2u(t)$? ($u(t)$ is the unit step function.)

(A) $-5/4$

(B) $0$

(C) $8/5$

(D) $5/4$

Problems 43–45 are based on the following statement.

A system has a transfer function given by

$$H(s) = \frac{s - B}{s^2 - 6s + A}$$

**43.** The system becomes unstable when a driving impulse with a frequency of 6 Hz is applied. What is the value of $A$.

(A) $-5$

(B) $-3$

(C) $-1$

(D) $0$

**44.** The system is stable when no impulse is applied. What is the value of $B$?

(A) $0$

(B) $6$

(C) $8$

(D) $10$

**45.** How many poles does the system have?

(A) $0$

(B) $1$

(C) $2$

(D) $3$

**46.** In the case of an ethical conflict, which entity takes precedence?

(A) society

(B) the client

(C) the employer

(D) society and the client are equal and take precedence

**47.** Who owns the rights to inventions that arise during work for a client if no prior specifications regarding inventions is made?

(A) the inventor

(B) the inventor's company

(C) the client

(D) the state

**48.** Which of the following is an ethics violation specifically included in the NCEES sample code of ethics?

(A) an engineering professor "moonlighting" as a private contractor

(B) an engineer investing money in the stock of the company for which he/she works

(C) a civil engineer with little electrical experience signing the plans for an electric generator

(D) none of the above

**49.** When may registrants coordinate entire projects?

(A) if each segment is signed and sealed by registrants responsible for the specific segments

(B) at no time

(C) with a special certification

(D) if all aspects of the project lie within the registrant's area of expertise

**50.** Which of the following is not an ethics violation?

(A) designing a product to fail in a specified number of years

(B) charging a premium for a superior product

(C) using proceeds from one product in order to sell another product below fair market price

(D) all of the above

**51.** When is it ethical to issue statements on technical issues inspired by interested parties?

(A) at no time

(B) if no compensation is received

(C) if there is no personal interest in the issue

(D) if the interested parties and their interests are explicitly revealed

**52.** To whom does a professional have ethical responsibilities?

(A) the employer

(B) consumers

(C) competitors

(D) all of the above

**53.** Criticism of another professional's work is

(A) never ethical

(B) ethical only if it directly relates to public safety and welfare

(C) ethical if truthful

(D) none of the above

**54.** Which entity passes engineering registration laws establishing the minimum criteria required to protect the public?

(A) state legislature

(B) professional societies

(C) national board of registration

(D) international registering councils

**55.** The cost of producing an item is represented by the following equation.

$$C_1 = \$25,\!000 + 0.03P$$

$P$ represents the number of items produced. If the items are sold for $1.50 each, approximately how many must be sold to break even?

(A) 1100

(B) 16,700

(C) 17,000

(D) 17,500

**56.** A factory is running at 80% efficiency with a fixed cost of $3000, a variable cost per unit of $5, a selling price per unit of $16, and a production capacity of 5000 units. What is the approximate current profit of the factory if all products manufactured are sold?

(A) $41,000

(B) $44,000

(C) $52,000

(D) $55,000

**57.** A proposal is made to buy a machine for $150,000. The expected service life is 15 yr with zero salvage value. What is most nearly the capitalized cost if the machine is kept in service indefinitely? The interest rate is 8%.

(A) $150,000

(B) $175,000

(C) $219,000

(D) $225,000

**58.** Approximately how much money should be deposited now in a savings account to yield $1000 in 6 yr? The annual interest rate is 10%, compounded semiannually.

(A) $459

(B) $557

(C) $565

(D) $679

**59.** A sum of money is deposited now in a savings account. The effective annual interest rate is 12%, and interest is compounded monthly. Approximately how much money must be deposited to yield $500 at the end of 11 mo?

    (A) $153
    (B) $446
    (C) $451
    (D) $500

**60.** A loan of $5000 is made for 5 yr at a simple interest rate of 8% per yr. How much total interest is paid?

    (A) $400
    (B) $2000
    (C) $5000
    (D) $7000

**61.** You decide to save a uniform amount at the end of each month for 12 months so you will have the $1000 at the end of 1 yr. The bank where you have a savings account pays 6% interest per annum compounded monthly. How much money do you need to deposit each month?

    (A) $70
    (B) $78
    (C) $81
    (D) $83

**62.** $9000 is to be invested now at 7% effective annual interest. A withdrawal is to be made at the end of each year for 10 yr. The 10 annual withdrawals will be equal, and the tenth will exhaust the fund. The amount of each withdrawal is most nearly

    (A) $1000
    (B) $1100
    (C) $1200
    (D) $1300

**63.** A sum of $60,000 will be needed for building improvements in 5 yr. To generate this amount, a sinking fund is established into which three equal payments will be made, one at the end of each of the first 3 yr. After the third year, no further payments will be made. If an effective annual interest rate of 8% can be expected, the amount that must be paid into the fund each year is most nearly

    (A) $16,000
    (B) $17,000
    (C) $19,000
    (D) $22,000

**64.** For the system of pulleys shown, most nearly what force, $P$, is required to lift the 1000 N load? Assume the pulleys are frictionless and weightless.

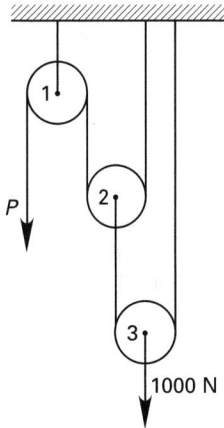

    (A) 125 N
    (B) 250 N
    (C) 330 N
    (D) 500 N

**65.** A cube of weight $W$ and side length $a$ is at rest on a flat surface with a friction coefficient of $\mu$. A cord attached to its top edge is pulled in a horizontal direction. The value of $\mu$ for which it is impossible to predict whether the cube will tip or slide is most nearly

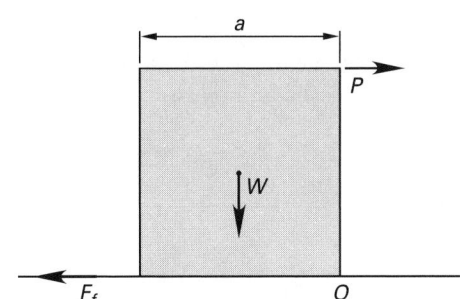

    (A) $\mu = \dfrac{P}{W}$

    (B) $\mu = \dfrac{W}{P}$

    (C) $\mu = 0.25$
    (D) $\mu = 0.5$

**66.** A 500 N force, $\mathbf{F}$, is directed as shown. What is most nearly the component vector representation of the force?

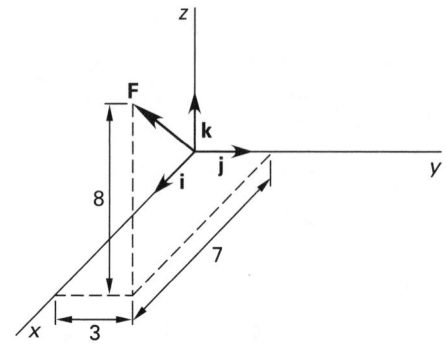

(A)  $(0.014 \text{ N})\mathbf{i} + (0.006 \text{ N})\mathbf{j} + (0.016 \text{ N})\mathbf{k}$
(B)  $(317 \text{ N})\mathbf{i} + (136 \text{ N})\mathbf{j} + (362 \text{ N})\mathbf{k}$
(C)  $(194 \text{ N})\mathbf{i} + (83 \text{ N})\mathbf{j} + (222 \text{ N})\mathbf{k}$
(D)  $(7 \text{ N})\mathbf{i} + (3 \text{ N})\mathbf{j} + (8 \text{ N})\mathbf{k}$

**67.** The change in the linear momentum of a particle is equivalent to which of the following?

(A)  the change in the particle's kinetic energy
(B)  the work performed on the particle
(C)  the impulse applied to the particle
(D)  the particle's mass times the distance the particle traveled

Problems 68 and 69 are based on the following statement.

A projectile is launched from a level plane at 30° from horizontal with an initial velocity of 1250 m/s.

**68.** What is most nearly the maximum height above the plane the projectile will reach?

(A)  20 km
(B)  40 km
(C)  60 km
(D)  80 km

**69.** What is most nearly the maximum range of the projectile?

(A)  40 km
(B)  70 km
(C)  140 km
(D)  160 km

Problems 70 and 71 are based on the following statement.

A 60 kg ball is dropped from a height of 48 m above a table.

**70.** What is most nearly the velocity of the ball just before impact?

(A)  11 m/s
(B)  15 m/s
(C)  22 m/s
(D)  31 m/s

**71.** If the coefficient of restitution between the ground and the ball is 0.9, what is most nearly the kinetic energy of the ball immediately after impact?

(A)  12 kJ
(B)  16 kJ
(C)  23 kJ
(D)  34 kJ

Problems 72 and 73 are based on the following statement and illustration.

A wheel with a radius of 2.5 m is pinned at point O, has two forces applied at its rim, and has a third force applied at a distance $R$ from the center. Additionally, a moment, $M$, of 200 N·m is applied at the center of the wheel.

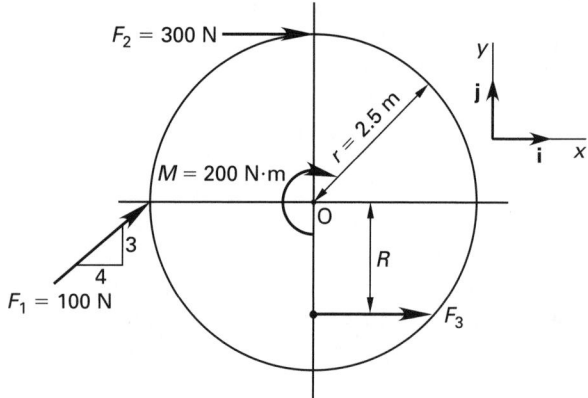

**72.** For equilibrium conditions, what is most nearly the distance $R$ if $F_3 = 500$ N?

(A)  1.4 m
(B)  1.6 m
(C)  1.8 m
(D)  2.2 m

**73.** The 200 N·m moment is removed and distance $R$ is 1.0 m. In order for the system to be in equilibrium, $F_3$ is most nearly

(A)  −500 N
(B)  −380 N
(C)  600 N
(D)  900 N

Problems 74 and 75 are based on the following statement.

At $t = 0$, a car traveling south is 10 km directly north of an eastbound bus. The car is traveling at 55 kph, and the bus is traveling at 25 kph. Visibility in all directions is 6 km.

**74.** Which equation gives the distance, $d$, in kilometers, between the car and the bus as a function of time, $t$, in hours?

(A) $d = 10 - 30t$
(B) $d = \sqrt{(10 + 55t)^2 + (25t)^2}$
(C) $d = \sqrt{100 - (60.4t)^2}$
(D) $d = \sqrt{3650t^2 - 1100t + 100}$

**75.** What is most nearly the velocity of the bus relative to the car at $t = 0$?

(A) 25 kph
(B) 49 kph
(C) 55 kph
(D) 60 kph

**76.** A rectangular beam 4 cm wide and 6 cm high is subjected to a shear of 7000 N at a particular location. The beam is constructed of 2014-T3 aluminum. What is most nearly the maximum shear stress at that location?

(A) $290 \text{ N/cm}^2$
(B) $440 \text{ N/cm}^2$
(C) $520 \text{ N/cm}^2$
(D) $660 \text{ N/cm}^2$

**77.** A 1020 carbon steel rod is $1/4$ cm in diameter and 6 cm long. The shear modulus is $11.5 \times 10^6 \text{ N/cm}^2$. Most nearly what torque must be applied to twist the rod 8°?

(A) 100 N·cm
(B) 120 N·cm
(C) 270 N·cm
(D) 420 N·cm

**78.** Two cantilever beams of equal length have dimensions of the beams are 0.5 cm × 0.5 cm and 0.25 cm × 0.25 cm, respectively. (Assume all other factors are equal.) The ratio of the end deflections, $y_{0.5}/y_{0.25}$, is most nearly

(A) 0.0625
(B) 0.125
(C) 0.500
(D) 1.25

**79.** Two rods are securely bonded at a common end and fixed to a wall. Rod A is made of aluminum ($E = 6.9 \times 10^6 \text{ N/cm}^2$) and has a diameter of 1.7 cm. Rod B is made of steel ($E = 20.7 \times 10^6 \text{ N/cm}^2$) and has a diameter of 0.8 cm. Both rods are initially 4 cm long. When a force, $P$, of magnitude 40 000 N is applied axially at the end of rod B, what is most nearly the total elongation experienced by the rods?

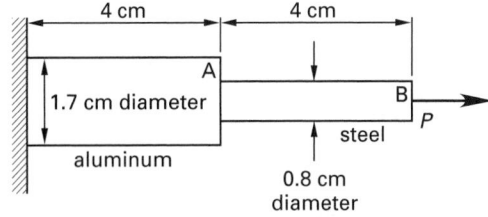

(A) 0.0063 cm
(B) 0.014 cm
(C) 0.026 cm
(D) 0.050 cm

**80.** A 1.5 cm × 0.75 cm steel bar is bent into the configuration shown. An axial load of 1500 N is applied to the ends. What is most nearly the maximum tensile stress of an element at section A-A?

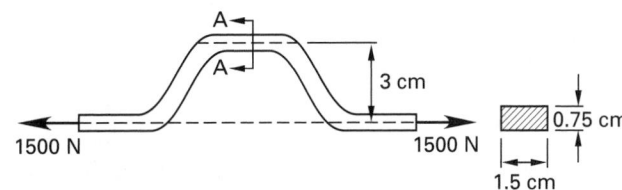

(A) $1.33 \text{ kN/cm}^2$
(B) $9.33 \text{ kN/cm}^2$
(C) $32.0 \text{ kN/cm}^2$
(D) $33.3 \text{ kN/cm}^2$

**81.** A steel plate ($E_s = 20 \times 10^6 \text{ N/cm}^2$) is securely bonded to a wooden beam ($E_w = 1 \times 10^6 \text{ N/cm}^2$). Most nearly where is the neutral axis with respect to the bottom of the composite beam base?

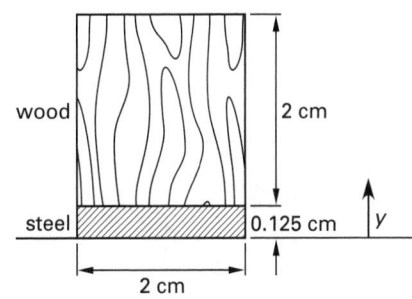

(A) 0.48 cm
(B) 0.54 cm
(C) 0.78 cm
(D) 0.96 cm

**82.** A cylindrical metal specimen with properties described by the following graph has a Poisson ratio of 0.3. What is the approximate change in diameter of the specimen if it is loaded axially to the elastic limit?

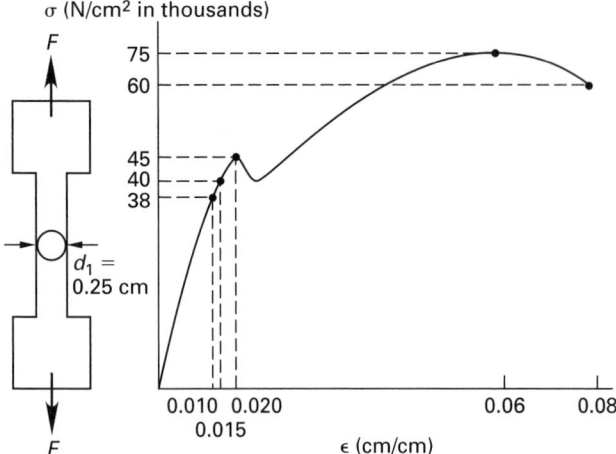

(A) 11 $\mu$m
(B) 15 $\mu$m
(C) 45 $\mu$m
(D) 60 $\mu$m

**83.** For a transversely loaded beam, the second derivative of the transverse deflection with respect to the axial distance along the beam is proportional to which of the following?

(A) shear
(B) moment
(C) load per unit length
(D) normal stress

**84.** What is the largest allowable axial load, $P$, that the fixed end column shown can bear without buckling, assuming a factor of safety for buckling of 3.0? The modulus of elasticity is $E = 200 \times 10^9$ Pa, and the centroidal moment of inertia is $I = 1.0 \times 10^{-4}$ m$^4$

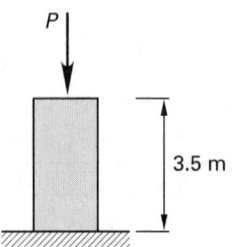

(A) 1100 kN
(B) 1209 kN
(C) 1343 kN
(D) 1390 kN

**85.** Which of the following is a synonym for electrovalent bonding?

(A) ionic bonding
(B) covalent bonding
(C) double bonding
(D) critical bonding

**86.** Which of the following factors will affect the hardenability of steel?

I.   composition
II.  grain size
III. lattice structure
IV.  cooling rate

(A) I only
(B) I and II
(C) II and III
(D) I, II, III, and IV

**87.** The yield strength of common yellow brass (70% Cu, 30% Zn) can be increased by which of the following?

(A) heat treatment
(B) annealing
(C) chill casting
(D) cold working

**88.** The elastic modulus, yield strength, ultimate tensile strength, and ductility of a metal can all be determined from

(A) an endurance test
(B) an impact test
(C) a quenching test
(D) a standard tensile test

**89.** Consider the stress-strain diagram for a carbon steel in tension. Determine the region where strain hardening occurs.

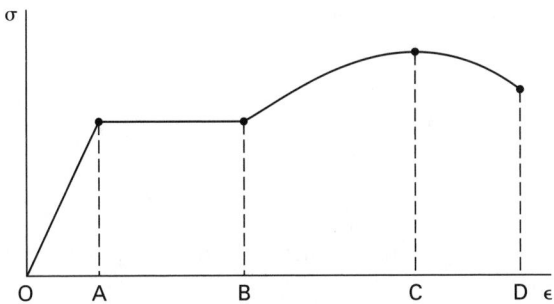

(A) O to A
(B) A to B
(C) B to C
(D) C to D

**90.** For alloys of two or more elements, Gibb's phase rule relates the number of degrees of freedom, $F$, to the number of phases, $P$, and the number of elements, $C$, in an equilibrium mixture. The rule is generally expressed as

(A) $P + C = F + 2$
(B) $P + C = 2F$
(C) $P + F = C + 2$
(D) $P = F + C + 2$

**91.** A 3 m long aluminum bar with a modulus of elasticity of 70 GPa is subjected to a tensile stress of 120 MPa. The elongation is most nearly

(A) 3.5 mm
(B) 5.1 mm
(C) 7.5 mm
(D) 9.0 mm

**92.** Which of the following may be the Poisson ratio of a material?

(A) −0.37
(B) 0.25
(C) 0.55
(D) 1.5

**93.** When a liquid flows under pressure through a pipe, the head loss due to surface friction with the pipe is $h_L = f(L/D)(\mathrm{v}^2/2g)$. Which of the following statements is false?

(A) The equation is valid for laminar as well as turbulent flow.
(B) The variable $D$ is the depth of flow in the pipe.
(C) The friction factor, $f$, is a function of a Reynolds number.
(D) The head loss, $h_L$, is expressed in units of distance.

**94.** What is most nearly the height of a column of carbon tetrachloride (specific gravity 1.56) that supports a pressure of 1 kPa?

(A) 0.0065 cm
(B) 6.5 cm
(C) 10 cm
(D) 64 cm

**95.** The velocity at a point on a model of a spillway for a dam is 5 m/s. If the length-to-scale ratio is 15:1, what is most nearly the velocity at the corresponding point on the actual dam? (Assume similar conditions.)

(A) 6.7 m/s
(B) 7.5 m/s
(C) 15 m/s
(D) 19 m/s

**96.** The transition between laminar and turbulent flow usually occurs at a Reynolds number of approximately

(A) 900
(B) 1200
(C) 1500
(D) 2100

**97.** A floating object is stable when the center of

(A) gravity is above the center of buoyancy
(B) buoyancy is above the center of gravity
(C) buoyancy is at the center of gravity
(D) gravity is above the metacenter

**98.** The theoretical fluid velocity through a nozzle generated by a 10 m hydraulic head is most nearly

(A) 4.5 m/s
(B) 9.9 m/s
(C) 14 m/s
(D) 200 m/s

**99.** Where does the vena contracta caused by a sharp-edged hydraulic orifice usually occur?

(A) at the centerline of the orifice
(B) at a distance of about 10% of the orifice diameter upstream from the plane of the orifice
(C) at a distance within 10% of the orifice diameter downstream from the plane of the orifice
(D) at a distance equal to about one-half the orifice diameter downstream from the plane of the orifice

**100.** A blower-type fan delivers air at the rate of 110 m³/min against a static gage pressure of 6 cm of water. Most nearly what net power is being delivered by the fan if the air has a density of 1.2 kg/m³? (Neglect velocity pressure.)

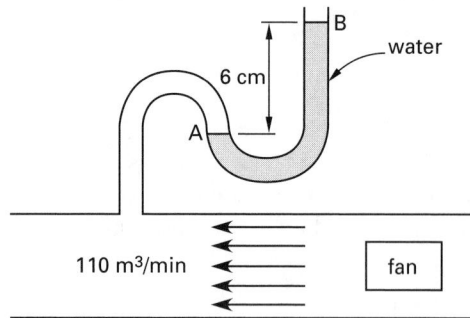

(A)  0.0016 W
(B)  110 W
(C)  1.1 kW
(D)  110 kW

**101.** The hydraulic radius of a 5 m deep triangular channel with a 1:1 side slope is most nearly

(A)  1.0 m
(B)  1.8 m
(C)  2.0 m
(D)  2.8 m

**102.** For A = 1, B = 0, E = 1, and F = 0, choose the following correct outputs.

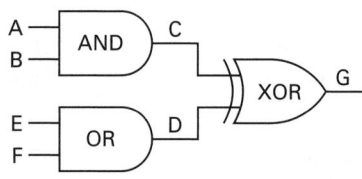

(A)  C = 1, D = 1, and G = 0
(B)  C = 1, D = 0, and G = 1
(C)  C = 0, D = 1, and G = 1
(D)  C = 0, D = 1, and G = 0

**103.** What is most nearly the current in the 6 Ω resistor?

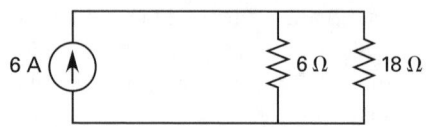

(A)  1.5 A
(B)  2 A
(C)  4 A
(D)  4.5 A

**104.** What is most nearly the equivalent capacitance between terminals A and B for the following circuit?

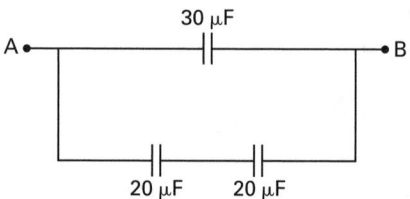

(A)  7.5 μF
(B)  32 μF
(C)  40 μF
(D)  70 μF

**105.** What is most nearly the resonant frequency of the circuit?

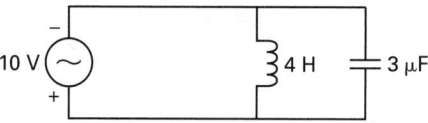

(A)  1.9 Hz
(B)  4.6 Hz
(C)  46 Hz
(D)  75 Hz

**106.** The power factor of a single-phase alternating-current circuit is defined as which of the following?

(A)  the ratio of apparent power (kVA) to real power (kW)
(B)  the ratio of real power (kW) to apparent power (kVA)
(C)  the ratio of real power to imaginary power
(D)  the ratio of reactive power to real power

**107.** What is most nearly the line-to-line voltage, $V_{ab}$, of the balanced wye connection?

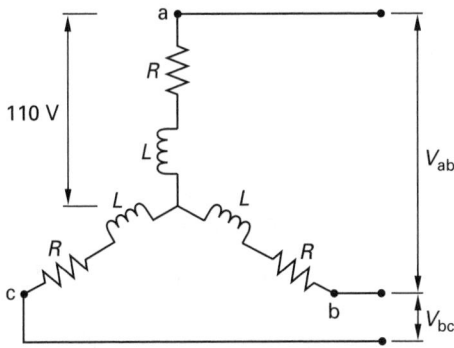

(A)  37 V
(B)  110 V
(C)  190 V
(D)  220 V

**108.**  What is most nearly the work required to move a charge of 10 C for a distance of 5 m in the same direction as a field of 50 V/m?

(A)  20 J
(B)  100 J
(C)  2.5 kJ
(D)  13 kJ

**109.**  If a passive parallel $RLC$ circuit is underdamped, the circuit's condition be changed to overdamped by

(A)  decreasing inductance ($L$)
(B)  decreasing the capacitance ($C$) and inductance ($L$)
(C)  increasing the resistance ($R$)
(D)  decreasing the resistance ($R$)

**110.**  For the circuit shown, what is the equivalent resistance between points a and b?

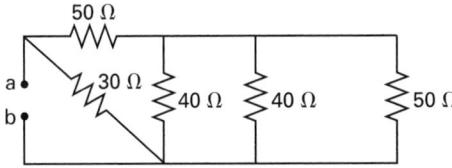

(A)  7 $\Omega$
(B)  20 $\Omega$
(C)  29 $\Omega$
(D)  50 $\Omega$

**111.**  Most nearly, how long would it take to charge the capacitor shown to 80% of the battery voltage in the circuit below, if the capacitor initially has no charge?

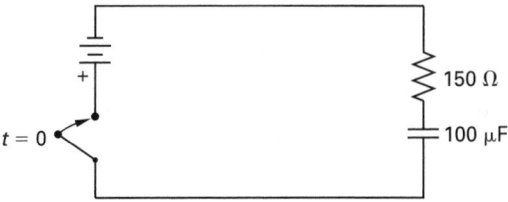

(A)  10 ms
(B)  15 ms
(C)  20 ms
(D)  24 ms

**112.**  A unit mass of steam is expanded in a cylinder in such a manner that no energy is added or lost as heat. What is this process called?

(A)  a constant enthalpy process
(B)  an isothermal process
(C)  an adiabatic process
(D)  a constant volume process

**113.**  Which of the following statements is true for a vapor dome drawn on a $T$-$s$ (temperature-entropy) diagram?

(A)  The bell-shaped curve indicates the saturation points for a constant temperature.
(B)  The region under the bell-shaped curve indicates superheating.
(C)  The bell-shaped curve indicates the saturation points for various pressures.
(D)  The left side of the bell-shaped curve indicates saturated vapor.

**114.**  One kg of air is stirred at a constant pressure of 1 atm so that the temperature increases from 500K to 600K. The stirring work is most nearly

(A)  3.4 kJ
(B)  28 kJ
(C)  41 kJ
(D)  100 kJ

**115.**  Use the information provided to determine the specific volume of ammonia at 278K and 10% quality.

$$v_{\text{saturated 278K liquid}} = 0.001\,58 \text{ m}^3/\text{kg}$$

$$v_{\text{saturated 278K vapor}} = 0.2479 \text{ m}^3/\text{kg}$$

(A)  0.0016 m$^3$/kg
(B)  0.026 m$^3$/kg
(C)  0.028 m$^3$/kg
(D)  0.25 m$^3$/kg

**116.** What is the approximate total heat transfer necessary to evaporate and superheat 1 kg of saturated liquid water in a boiler? The water is initially at 1 atm and remains so throughout the evaporation process. The final temperature of the vapor is 600°C.

(A)  3.3 MJ
(B)  6.2 MJ
(C)  7.2 MJ
(D)  8.1 MJ

**117.** Which group contains processes that are always part of any physically realizable vapor power cycle?

(A)  isentropic compression and isentropic expansion
(B)  isentropic compression
(C)  adiabatic heat addition and adiabatic heat extraction
(D)  isobaric vaporization and isobaric condensation

**118.** 210 m$^3$ of water passes through a heat exchanger and absorbs 26 400 MJ of heat energy. The exit temperature is 48°C. The water density is 1000 kg/m$^3$. Calculate the entrance temperature.

(A)  12°C
(B)  18°C
(C)  30°C
(D)  48°C

**119.** Which of the following is an extensive property?

(A)  pressure
(B)  specific enthalpy
(C)  temperature
(D)  internal energy

**120.** Which of the following statements is true?

(A)  Equal volumes of different gases have equal weights.
(B)  High-density gases diffuse faster than low-density gases.
(C)  Equal volumes of different gases at the same temperature and pressure have an equal number of molecules.
(D)  Equal volumes of different gases at the same temperature and pressure have equal densities.

# STOP!

## DO NOT CONTINUE ON.

This concludes the Morning Section of the examination. If you finish before time is called, you may check your work on this section of the examination. You may not turn to the Afternoon Section of the exam until you are told to do so by your proctor. Be sure that all of your responses on the answer sheet are dark and completely fill the bubbles.

# Instructions

This section of the exam consists of 60 problems, each worth 2 points. You will have four hours in which to work this section. Your score will be determined by the number of problems that you solve correctly. No points will be deducted for incorrect answers. It is to your best advantage to try to answer every question.

When permission has been given by your proctor, break the seal on the Examination Booklet and remove the Answer Sheet. Write your name immediately in the space indicated. Check that all pages are present and legible. If any part of this Booklet is missing, your proctor will issue you a new Booklet.

All solutions must be entered on the Answer Sheet. No credit will be given for answers appearing only in the Examination Booklet. Mark your answers with the pencil provided. Marks must be dark and must completely fill the bubble. Record only one answer per problem; if you mark more than one answer, you will not receive credit for the problem. If you change an answer, be sure the old bubble is erased completely; incomplete erasures may be read as intended answers.

If you finish early, check your work and make sure you have correctly followed all instructions. After checking your answers, you may turn in your Examination Booklet and Answer Sheet and leave the examination room. Once you leave, you will not be permitted to return to work on your solutions.

Do not work any problems from the Afternoon Section of the exam during the first four hours of this exam.

WAIT FOR PERMISSION TO BEGIN.

Name: _____

| Last | First | Middle |
| | | Initial |

## FUNDAMENTALS OF ENGINEERING SAMPLE EXAMINATION

### EXAM 2
### AFTERNOON SECTION

#### Subject Breakdown

The major subject areas and their corresponding problem numbers are listed below.

| | |
|---|---|
| Advanced Engineering Mathematics | 1–6 |
| Engineering Probability and Statistics | 7–11 |
| Biology | 12–14 |
| Engineering Economics | 15–20 |
| Application of Engineering Mechanics | 21–28 |
| Engineering of Materials | 29–35 |
| Fluids | 36–44 |
| Electricity and Magnetism | 45–51 |
| Thermodynamics and Heat Transfer | 52–60 |

# Fundamentals of Engineering Sample Examination

## Afternoon Section

Name: _____

| | | | | |
|---|---|---|---|---|
| 1. (A) (B) (C) (D) | 16. (A) (B) (C) (D) | 31. (A) (B) (C) (D) | 46. (A) (B) (C) (D) |
| 2. (A) (B) (C) (D) | 17. (A) (B) (C) (D) | 32. (A) (B) (C) (D) | 47. (A) (B) (C) (D) |
| 3. (A) (B) (C) (D) | 18. (A) (B) (C) (D) | 33. (A) (B) (C) (D) | 48. (A) (B) (C) (D) |
| 4. (A) (B) (C) (D) | 19. (A) (B) (C) (D) | 34. (A) (B) (C) (D) | 49. (A) (B) (C) (D) |
| 5. (A) (B) (C) (D) | 20. (A) (B) (C) (D) | 35. (A) (B) (C) (D) | 50. (A) (B) (C) (D) |
| 6. (A) (B) (C) (D) | 21. (A) (B) (C) (D) | 36. (A) (B) (C) (D) | 51. (A) (B) (C) (D) |
| 7. (A) (B) (C) (D) | 22. (A) (B) (C) (D) | 37. (A) (B) (C) (D) | 52. (A) (B) (C) (D) |
| 8. (A) (B) (C) (D) | 23. (A) (B) (C) (D) | 38. (A) (B) (C) (D) | 53. (A) (B) (C) (D) |
| 9. (A) (B) (C) (D) | 24. (A) (B) (C) (D) | 39. (A) (B) (C) (D) | 54. (A) (B) (C) (D) |
| 10. (A) (B) (C) (D) | 25. (A) (B) (C) (D) | 40. (A) (B) (C) (D) | 55. (A) (B) (C) (D) |
| 11. (A) (B) (C) (D) | 26. (A) (B) (C) (D) | 41. (A) (B) (C) (D) | 56. (A) (B) (C) (D) |
| 12. (A) (B) (C) (D) | 27. (A) (B) (C) (D) | 42. (A) (B) (C) (D) | 57. (A) (B) (C) (D) |
| 13. (A) (B) (C) (D) | 28. (A) (B) (C) (D) | 43. (A) (B) (C) (D) | 58. (A) (B) (C) (D) |
| 14. (A) (B) (C) (D) | 29. (A) (B) (C) (D) | 44. (A) (B) (C) (D) | 59. (A) (B) (C) (D) |
| 15. (A) (B) (C) (D) | 30. (A) (B) (C) (D) | 45. (A) (B) (C) (D) | 60. (A) (B) (C) (D) |

**1.** What is the equation for the line tangent to the parabola at point $(2, 40/9)$?

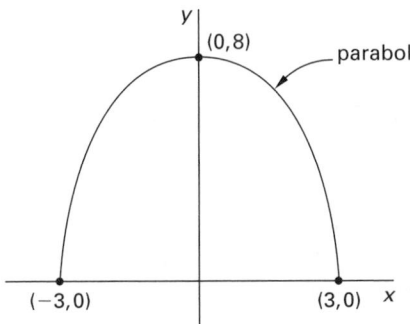

(A) $y = -\dfrac{16}{9}x + 8$

(B) $x = -\dfrac{9}{32}y + \dfrac{3}{4}$

(C) $x = -\dfrac{9}{32}y + \dfrac{13}{4}$

(D) $y = -\dfrac{32}{9}x + \dfrac{8}{3}$

Problems 2 and 3 are based on the following statement.

Radioactive $C^{14}$ undergoes exponential decay with a half-life of 5730 years.

**2.** What is most nearly the decay constant of $C^{14}$?

(A) $8.7 \times 10^{-5}/\text{yr}$
(B) $1.2 \times 10^{-4}/\text{yr}$
(C) $1.8 \times 10^{-4}/\text{yr}$
(D) $2.6 \times 10^{-3}/\text{yr}$

**3.** In a given sample of $C^{14}$, most nearly what percentage will have decayed after 10,000 years?

(A) 17%
(B) 30%
(C) 70%
(D) 83%

**4.** Where is the centroid of the area bounded by the curve $y = x^3$, the $x$-axis, and the lines $x = 0$ and $x = 3$?

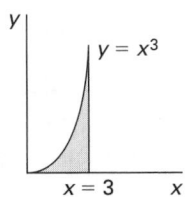

(A) $(\overline{x}, \overline{y}) = (1.6, 7.8)$
(B) $(\overline{x}, \overline{y}) = (1.8, 5.8)$
(C) $(\overline{x}, \overline{y}) = (2.0, 18.0)$
(D) $(\overline{x}, \overline{y}) = (2.4, 7.7)$

Problems 5 and 6 are based on the following equation.

$$y = 3x^3 - 3x^2 - 24x + 25$$

**5.** What is most nearly the slope of the curve, $y$, at the point $(-3, -11)$?

(A) 3.0
(B) 39
(C) 75
(D) 100

**6.** At most nearly which value of $y$ does the relative minimum occur?

(A) $-19.5$
(B) $-11.0$
(C) 37.0
(D) 44.6

**7.** The least-squares method is used to plot a straight line through the data points $(8,-1)$, $(2,2)$, $(-3,6)$, and $(-9,11)$. The correlation coefficient is most nearly

(A) $-0.87$
(B) $-0.91$
(C) $-0.95$
(D) $-0.99$

**8.** On average, 30% of a certain business's customers are women. Ten customers are chosen at random to be surveyed. The probability that exactly three of them are women is most nearly

(A) 0.16
(B) 0.27
(C) 0.33
(D) 0.42

**9.** The best curve of the form $y = a + b\sqrt{x}$ is fitted to the points $(2,8)$, $(5,11)$, $(10,15)$, and $(21,20)$. The value of $b$ is most nearly

(A) 3.8
(B) 4.2
(C) 5.0
(D) 5.8

**10.** When operating properly, a plant has a daily production rate that is normally distributed with a mean of 144 kg/d and a standard deviation of 25 kg/d. During an analysis period, the output is measured on 30 consecutive days, and the mean output is found to be 135 kg/d. The probability that the plant is not operating properly is most nearly

(A) 1%
(B) 5%
(C) 95%
(D) 99%

**11.** The reliability of a machine is exponentially distributed with a mean time to failure of 650 h. The probability that the machine will not have failed after 750 h of operation is most nearly

(A) 0.054
(B) 0.12
(C) 0.28
(D) 0.32

**12.** A community water supply well was found to be contaminated with copper cyanide (CuCN), methanol (CH$_3$OH), and potassium cyanide (KCN). The concentrations in the water and the EPA oral reference doses are as follows.

| toxicant | exposure ($\mu$g/L) | RfD (oral) ($\mu$g/kg·d) |
|---|---|---|
| CuCN | 40 | 5 |
| CH$_3$OH | 1000 | 500 |
| KCN | 600 | 50 |

What is the sum of the hazard ratios for a 70 kg person who drinks 2 L of water daily for these noncarcinogens?

(A) 0.23
(B) 0.57
(C) 0.63
(D) 1.1

**13.** Workers are crushing and grading rock for a drainfield in a subsurface disposal system. Samples of the air in the breathing zone of the workers show the following.

| time | dust concentration (mg/m$^3$) | percent SiO$_2$ |
|---|---|---|
| 0800 | 1.5 | 5.8 |
| 1000 | 1.1 | 6.3 |
| 1300 | 2.1 | 4.2 |
| 1500 | 1.6 | 3.9 |
| 1700 | 1.8 | 4.1 |

To calculate the PELs (permissible exposure limits) for mixtures containing free silica, use

$$PEL = \frac{10 \ \frac{mg}{m^3}}{\%SiO_2 + 2}$$

At which time(s) did worker exposure exceed the OSHA PEL for quartz?

(A) 1300 hours
(B) 1000 and 1300 hours
(C) 1300, 1500, and 1700 hours
(D) 0800, 1300, and 1700 hours

**14.** Passive diffusion is influenced by which of the following?

I.   lipid solubility
II.  participation of a protein carrier molecule
III. molecular size
IV.  ionization

(A) II only
(B) I and III
(C) II and III
(D) I, III, and IV

Problems 15–17 are based on the following statement.

A $25,000 bank loan is to be repaid in equal yearly payments over 15 yr at an effective annual interest rate of 7%.

**15.** What is most nearly the yearly payment?

(A) $1667
(B) $1783
(C) $2745
(D) $3417

**16.** Approximately what percentage of the first payment is applied to the principal?

(A) 0%
(B) 36%
(C) 39%
(D) 51%

**17.** Assume the borrower can earn an annual effective return of 10% in the stock market and uses the loan money to do so. All money earned is kept in the stock fund. What is most nearly the present-day value of the earnings over 15 yr as a result of this loan and investment venture?

(A) $750
(B) $4121
(C) $8487
(D) $9325

**18.** A machine has an initial cost of $40,000 and operating costs of $3500 each year. Its salvage value decreases by $4000 each year. The machine is now 4 yr old. Assuming an effective annual interest rate of 12%, the cost of owning and operating the machine for one more year is most nearly

(A) $7500
(B) $8800
(C) $10,000
(D) $12,000

**19.** A machine has an initial cost of $23,000, a life of 10 yr, and a salvage value of $1800. The straight line depreciation value of this machine is most nearly

(A) $1400
(B) $1600
(C) $1800
(D) $2100

**20.** A machine has an initial cost of $15,000 and an annual maintenance cost of $550. The life of the machine is 17 yr and its salvage value is $3000. Assuming an effective annual interest rate of 6%, the equivalent uniform annual cost for the machine is most nearly

(A) $1500
(B) $1700
(C) $1900
(D) $2200

Problems 21 and 22 are based on the following statement and illustration.

In the truss shown, all members are pin-connected beams.

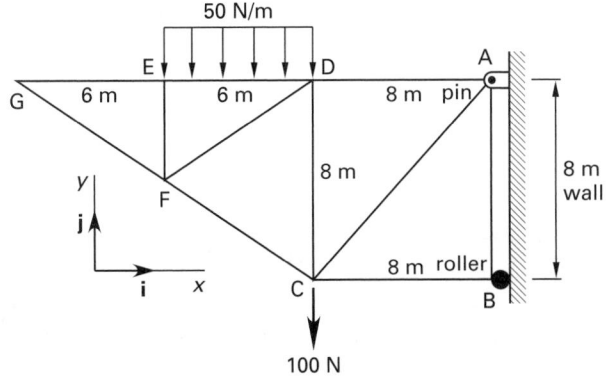

**21.** What are most nearly the reactions at A and B?

(A) $\mathbf{R}_A = (-512.5 \text{ N})\mathbf{i} + (400 \text{ N})\mathbf{j}$
    $\mathbf{R}_B = (512.5 \text{ N})\mathbf{i}$
(B) $\mathbf{R}_A = (412.5 \text{ N})\mathbf{i} + (400 \text{ N})\mathbf{j}$
    $\mathbf{R}_B = (-412.5 \text{ N})\mathbf{i}$
(C) $\mathbf{R}_A = (512.5 \text{ N})\mathbf{i} + (200 \text{ N})\mathbf{j}$
    $\mathbf{R}_B = (-512.5 \text{ N})\mathbf{i} + (200 \text{ N})\mathbf{j}$
(D) $\mathbf{R}_A = (512.5 \text{ N})\mathbf{i} + (400 \text{ N})\mathbf{j}$
    $\mathbf{R}_B = (-512.5 \text{ N})\mathbf{i}$

**22.** What is most nearly the force in member EF?

(A) 0 N
(B) 100 N
(C) 150 N
(D) 300 N

**23.** A hydraulic jack has a ram diameter at E of 1.5 cm and a pump piston diameter at D of 0.625 cm. The jack handle has a mechanical advantage of 15 to 1.

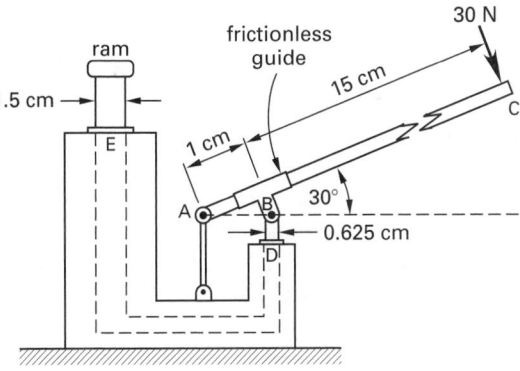

Neglecting friction, most nearly what maximum force can the jack produce when a 30 N force is applied to the handle at C in the position shown?

(A) 450 N
(B) 600 N
(C) 2400 N
(D) 2800 N

**24.** The truss is supported at A by a pin and at B by a roller. All members are pin-connected.

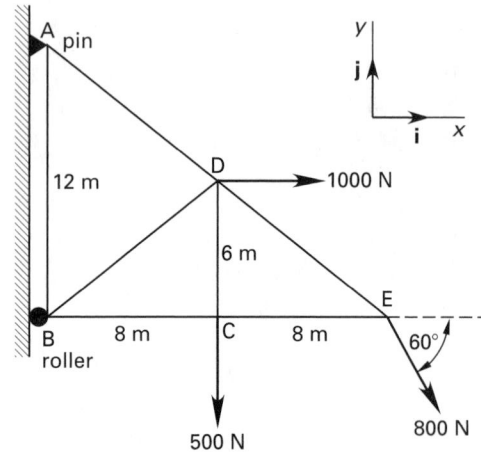

What are most nearly the reactions at A and B?

(A) $\mathbf{R}_A = (-1757.1\ \text{N})\mathbf{i} + (1192.8\ \text{N})\mathbf{j}$
    $\mathbf{R}_B = (357.1\ \text{N})\mathbf{i}$
(B) $\mathbf{R}_A = (1757.1\ \text{N})\mathbf{i} + (1192.8\ \text{N})\mathbf{j}$
    $\mathbf{R}_B = (-357.1\ \text{N})\mathbf{i}$
(C) $\mathbf{R}_A = (-1366.6\ \text{N})\mathbf{i} + (900\ \text{N})\mathbf{j}$
    $\mathbf{R}_B = (-326.2\ \text{N})\mathbf{i}$
(D) $\mathbf{R}_A = (-800.5\ \text{N})\mathbf{i} + (1192.8\ \text{N})\mathbf{j}$
    $\mathbf{R}_B = (-690.4\ \text{N})\mathbf{i}$

Problems 25–28 are based on the following statement and illustration.

One end of a 2 m long steel rod ($d = 3$ cm) is rigidly attached to a wall so that the rod is perpendicular to the wall. The other end of the rod is welded to a second rod, which is 1 m long ($d = 1.0$ cm). The second rod is parallel to the wall and the floor. A vertical force of 50 N is applied to the end of the second rod. For both rods, the modulus of elasticity is $210 \times 10^9$ Pa and the shear modulus is $79 \times 10^9$ Pa.

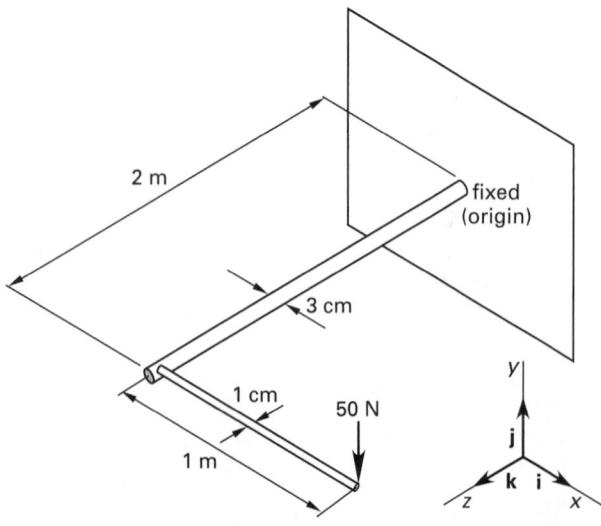

**25.** What are most nearly the reaction and moment at the wall?

(A) $\mathbf{R} = (50\ \text{N})\mathbf{j}$
    $\mathbf{M} = (-200\ \text{N·m})\mathbf{i}$
(B) $\mathbf{R} = (50\ \text{N})\mathbf{j}$
    $\mathbf{M} = (-100\ \text{N·m})\mathbf{k}$
(C) $\mathbf{R} = (50\ \text{N})\mathbf{j}$
    $\mathbf{M} = (200\ \text{N·m})\mathbf{i} + (-100\ \text{N·m})\mathbf{k}$
(D) $\mathbf{R} = (50\ \text{N})\mathbf{j}$
    $\mathbf{M} = -(100\ \text{N·m})\mathbf{i} + (-50\ \text{N·m})\mathbf{k}$

**26.** What is most nearly the maximum angle of twist for the 3 cm diameter rod?

(A) 0.91°
(B) 1.4°
(C) 1.8°
(D) 2.5°

**27.** What is most nearly the vertical (downward) deflection at the tip of the 3 cm diameter rod?

(A) 0.31 mm
(B) 1.6 mm
(C) 2.7 mm
(D) 1.6 cm

**28.** There is an infinitesimally small cubic element at the top surface of the 3 cm diameter rod immediately adjacent to the wall, that is, at $(0, 1.5, 0)$. What is most nearly the maximum normal stress due to the bending moment?

(A) 0 MPa
(B) 19 MPa
(C) 38 MPa
(D) 75 MPa

Problems 29–31 are based on the following statement and illustration.

The simplified phase diagram of an alloy of components A and B is shown.

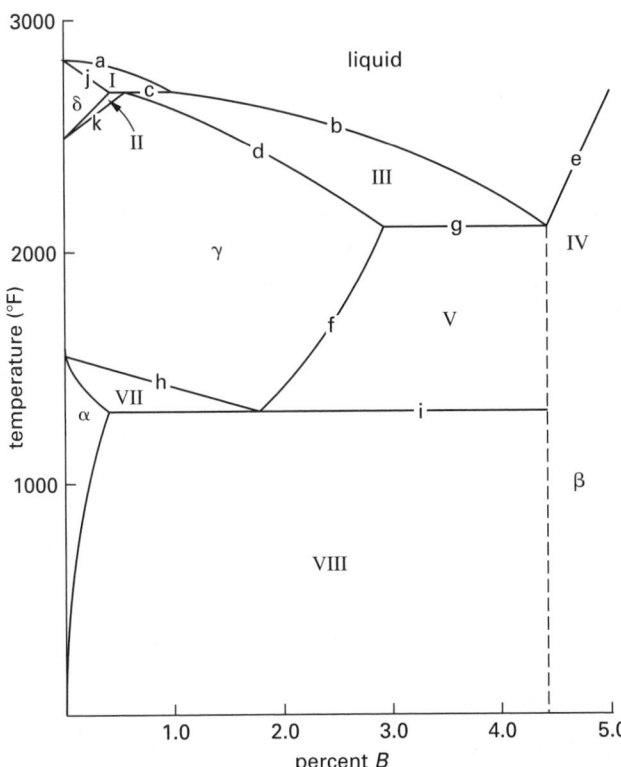

(A) 15 cm$^2$
(B) 30 cm$^2$
(C) 150 cm$^2$
(D) 300 cm$^2$

**33.** Which of the following is an elastoplastic material?

(A) plastic
(B) metal
(C) rubber
(D) glass

**34.** A material with a high elastic modulus, high yield point, high elongation at fracture, and high ultimate strength would exhibit which of the following curves on the stress-strain diagram?

(A)
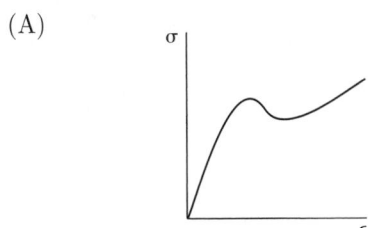

(B)
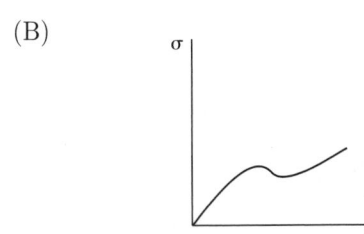

(C)
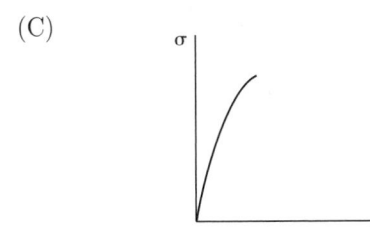

(D)
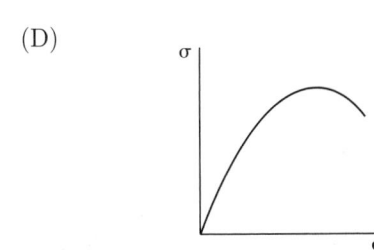

**29.** Which regions contain a mixture of liquid and solid phases?

(A) I and IV
(B) III only
(C) I, III, and IV
(D) I and III

**30.** What is the approximate percentage of solid alloy that will be present at 2300°F if the mixture is 3% B and 97% A?

(A) 49%
(B) 57%
(C) 82%
(D) 97%

**31.** What region(s) represent a mixture of the δ and γ constituents?

(A) II only
(B) III only
(C) II and VII
(D) VIII only

**32.** A 4 m long, 150 kg cylindrical metal bar is suspended vertically from one end. The metal has a density of 2500 kg/m$^3$ and modulus of elasticity of 210 GPa. The bar's elongation is $9.0 \times 10^{-7}$ m. What is most nearly the bar's cross-sectional area?

**35.** Which of the following statements is true?

(A) Low-alloy steels are a minor group and are rarely used.

(B) There are three basic types of stainless steels: martensitic, austenitic, and ferritic.

(C) The addition of small amounts of silicon to steel can cause a marked decrease in the yield strength of steel.

(D) The addition of small amounts of molybdenum to low alloy steels makes it possible to harden and strengthen thick pieces of the metal by heat treatment.

**36.** Approximately what depth of water, $h$, will produce a horizontal force of 2.5 N against the 2 cm × 2 cm plate?

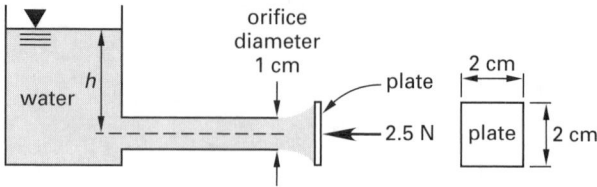

(A) 0.91 m
(B) 1.6 m
(C) 32 m
(D) 65 m

**37.** What is most nearly the total force acting on a 1 m wide section of the curved surface?

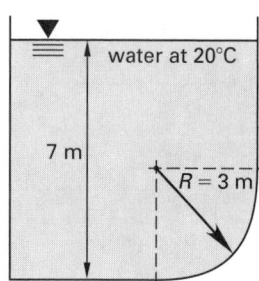

(A) 120 kN
(B) 160 kN
(C) 220 kN
(D) 250 kN

**38.** A venturi meter installed in a pipe with a 38.1 cm diameter has a throat diameter of 21.24 cm. The static gage pressure upstream of the venturi is 172.4 kPa. The average fluid velocity in the pipe is 7.62 m/s. The fluid flowing is water. If cavitation is just beginning at the throat of the venturi, what is most nearly the absolute vapor pressure of the water at the throat?

(A) 2.2 kPa
(B) 49 kPa
(C) 270 kPa
(D) 290 kPa

**39.** A 100 m surface ocean vessel is designed to travel at 110 kph. Approximately how fast must water flow over the hull of a $^1/_{25}$-scale model to be similar to the forces on the full-scale ship?

(A) 1.2 m/s
(B) 6.1 m/s
(C) 13 m/s
(D) 15 m/s

**40.** Water flows at 0.07 m³/s in a 0.5 m (inside diameter) sewer line with a Manning roughness coefficient, $n$, of 0.015 and a geometric slope, $S$, of 0.001. The Manning's coefficient varies with depth, and the flow is uniform and steady. The circular channel velocity ratio, v to $v_{full}$, is 0.94. The velocity is most nearly

(A) 0.50 m/s
(B) 0.34 m/s
(C) 0.23 m/s
(D) 0.10 m/s

**41.** Water flows at 14 m³/s in a 6 m wide rectangular channel. The critical velocity is most nearly

(A) 0.82 m/s
(B) 1.8 m/s
(C) 2.8 m/s
(D) 14 m/s

**42.** A trapezoidal channel with a depth of 7 m, a bottom width of 5 m, and 2:1 (horizontal:vertical) slopes, carries a water flow of 7 m³/s. The channel is lined with concrete ($n = 0.012$) and has a slope of 0.003. The flow velocity is most nearly

(A) 0.05 m/s
(B) 1.1 m/s
(C) 8.0 m/s
(D) 24 m/s

**43.** Water is flowing through a pipe. A pitot-static gauge registers 0.076 m of mercury ($\rho_m = 13\,580$ kg/m³). The velocity of water in the pipe is most nearly

(A) 1.3 m/s
(B) 2.2 m/s
(C) 3.8 m/s
(D) 4.3 m/s

**44.** Two parcels of land contribute runoff through a drainage channel to a detention pond. The storm intensity after 25 minutes is 3.9 m/h. The peak flow is most nearly

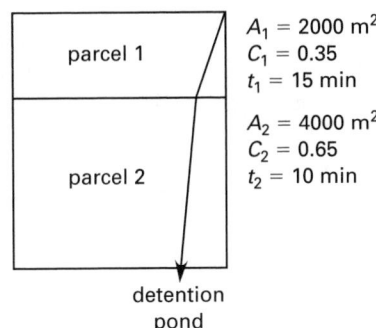

$A_1 = 2000 \text{ m}^2$
$C_1 = 0.35$
$t_1 = 15 \text{ min}$

$A_2 = 4000 \text{ m}^2$
$C_2 = 0.65$
$t_2 = 10 \text{ min}$

parcel 1

parcel 2

detention pond

(A) 1.0 m³/s
(B) 4.0 m³/s
(C) 6.0 m³/s
(D) 9.0 m³/s

Problems 45–47 are based on the following illustration.

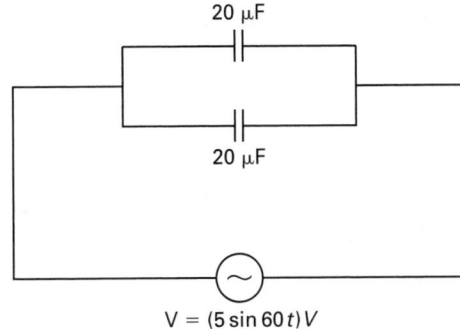

20 μF

20 μF

$V = (5\sin 60\,t)\,V$

**45.** What is most nearly the equivalent capacitance for the circuit?

(A) 0.1 μF
(B) 10 μF
(C) 20 μF
(D) 40 μF

**46.** What is most nearly the effective value of the current flowing across the capacitive combination?

(A) 0.021 mA
(B) 2.1 mA
(C) 8.5 mA
(D) 12 mA

**47.** What is most nearly the average power dissipated across the capacitive combination?

(A) 0
(B) 0.050 mW
(C) 0.50 mW
(D) 42 mW

**48.** What is the Thevenin equivalent for the following circuit?

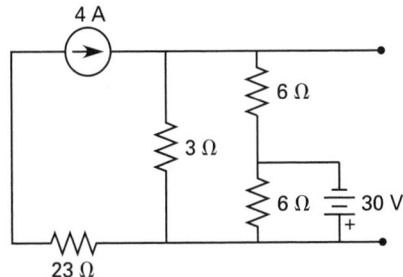

4 A

6 Ω

3 Ω

6 Ω

30 V

23 Ω

(A)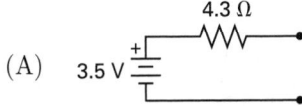

4.3 Ω

3.5 V

(B)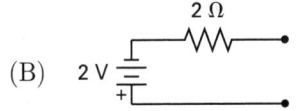

2 Ω

2 V

(C)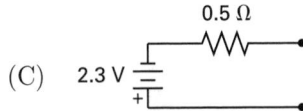

0.5 Ω

2.3 V

(D)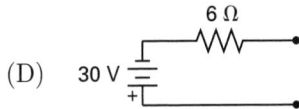

6 Ω

30 V

**49.** A circuit draws 5000 kVA with a power factor of 0.72. Most nearly what size capacitor is required to increase the power factor to 0.86? The 60 Hz line voltage is 220 V (rms).

(A) 0.20 mF
(B) 40 mF
(C) 70 mF
(D) 160 mF

**50.** A 240 V alternating source at 60 cycles is connected to a series $RLC$ circuit with values as shown below.

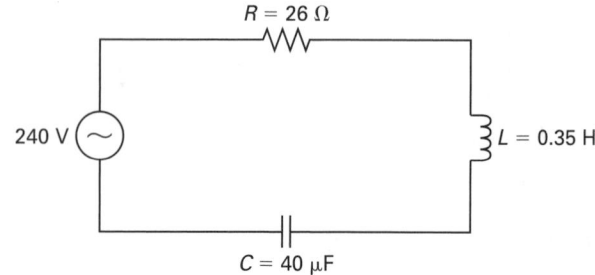

What is most nearly the total reactance of the series induction and capacitor?

(A)  66 Ω
(B)  130 Ω
(C)  150 Ω
(D)  200 Ω

**51.** The magnetic saturation curve limits the line voltage at which a generator or motor can operate. Which of the following statements is FALSE regarding saturation curves?

(A)  Saturation does not depend upon the type of steel used in the poles.
(B)  As field current increases, the hysteresis effect limits the increase in the flux produced.
(C)  Poles that allow the production of more flux permit higher operating voltages.
(D)  More flux at a constant field voltage can be produced by increasing the number of poles.

**52.** A gas at 1 atm pressure originally occupies a volume of 0.566 m$^3$. The gas is polytropically compressed with $n = 1.15$ until the pressure is 334.5 kPa. Most nearly what work is required for this compression?

(A)  61 J
(B)  160 J
(C)  410 J
(D)  64 kJ

**53.** One kilogram of air goes through the following cycle: From a pressure of 1 atm and a temperature of 20°C, the air is heated to 200°C at constant volume. It is then allowed to expand isentropically, performing useful work in the process, until the pressure is 1 atm. Finally, it is cooled to 20°C at constant pressure. What is most nearly the thermal efficiency of the cycle?

$$R = 287 \text{ J/kg·K}$$
$$c_p = 1000 \text{ J/kg·K}$$
$$c_v = 718 \text{ J/kg·K}$$
$$k = 1.4$$

(A)  7.8%
(B)  27%
(C)  34%
(D)  39%

**54.** An ideal gas at a pressure of 3.45 MPa and a temperature of 20°C is contained in a cylinder with a volume of 20 m$^3$. Enough gas is released to lower the pressure in the cylinder to 1.725 MPa. The expansion of the gas is isentropic. The specific heat ratio is 1.4, and the gas constant is 273 J/kg·K. Most nearly how much gas is released from the cylinder?

(A)  340 kg
(B)  510 kg
(C)  860 kg
(D)  1240 kg

**55.** Methane gas, CH$_4$, burns in air and releases 802 kJ/mol. Assuming a 90%-efficient heat transfer, most nearly what mass of ice at $-17.8$°C (latent heat of fusion of 333 kJ/kg) can be converted to water at 37.8°C by burning 500 L of methane measured at 20.56°C and 1 atm?

(A)  15 kg
(B)  20 kg
(C)  28 kg
(D)  130 kg

Problems 56 and 57 are based on the following statement and illustration.

A rigid, adiabatic box is partitioned by a moveable, adiabatic piston. On either side of the piston are two perfect gases, A and B. The system is initially in equilibrium with $P_A = P_B = 1$ atm and $T_A = T_B = 300$K. Gas B occupies 1 m$^3$. The valve is opened, and additional gas A flows into the left section until $P_A = 5$ atm. The ratio of specific heats for gas B is 1.4, and gas B has a specific heat, $C_v$, of 20 J/mol·K.

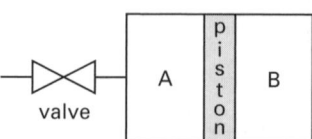

**56.** What is most nearly the final volume occupied by gas B?

(A)  0.11 m$^3$
(B)  0.20 m$^3$
(C)  0.28 m$^3$
(D)  0.32 m$^3$

**57.** Most nearly how much work is done on gas B?

(A) 3.5 kJ
(B) 24 kJ
(C) 52 kJ
(D) 140 kJ

**58.** 200 Btu of energy are used to heat 2 kg of water $(c_v = 4.2 \text{ kJ/kg·K})$. The change in temperature is most nearly

(A) 10 K
(B) 25 K
(C) 29 K
(D) 45 K

**59.** 40 g of oxygen gas ($O_2$) are compressed at a constant temperature of 20°C to 60% of the original volume. The universal gas constant in this system is 1.98 cal/gmol·K. Most nearly what work is done on the system?

(A) 210 cal
(B) 240 cal
(C) 370 cal
(D) 1200 cal

**60.** Which of the following statements is FALSE?

(A) The Carnot cycle is dependent on the source and sink temperatures, not the working fluid.
(B) The thermal efficiency of a power cycle is defined as the ratio of useful work output to the supplied input energy.
(C) The maximum work obtained from a cycle is dependent on the temperature of the local environment.
(D) Maximum work output will be obtained in an irreversible process.

# Exam 1—Answer Key

## Morning Section

| | | | | | |
|---|---|---|---|---|---|
| 1. C | 21. D | 41. D | 61. C | 81. D | 101. A |
| 2. A | 22. C | 42. B | 62. D | 82. D | 102. B |
| 3. C | 23. B | 43. A | 63. C | 83. C | 103. B |
| 4. C | 24. B | 44. D | 64. D | 84. C | 104. A |
| 5. C | 25. A | 45. B | 65. D | 85. A | 105. A |
| 6. C | 26. B | 46. C | 66. A | 86. C | 106. D |
| 7. B | 27. D | 47. D | 67. C | 87. A | 107. D |
| 8. A | 28. C | 48. B | 68. D | 88. A | 108. B |
| 9. C | 29. D | 49. B | 69. C | 89. B | 109. A |
| 10. C | 30. B | 50. A | 70. D | 90. D | 110. B |
| 11. D | 31. D | 51. D | 71. A | 91. D | 111. A |
| 12. C | 32. D | 52. A | 72. C | 92. C | 112. D |
| 13. C | 33. D | 53. B | 73. B | 93. A | 113. B |
| 14. D | 34. C | 54. B | 74. B | 94. C | 114. D |
| 15. D | 35. C | 55. D | 75. C | 95. D | 115. C |
| 16. A | 36. B | 56. B | 76. B | 96. B | 116. A |
| 17. D | 37. B | 57. D | 77. C | 97. C | 117. A |
| 18. C | 38. B | 58. B | 78. C | 98. B | 118. C |
| 19. C | 39. A | 59. B | 79. A | 99. C | 119. A |
| 20. A | 40. B | 60. B | 80. C | 100. C | 120. B |

## Afternoon Section

| | | | | | |
|---|---|---|---|---|---|
| 1. D | 11. D | 21. B | 31. B | 41. C | 51. B |
| 2. B | 12. B | 22. C | 32. C | 42. D | 52. D |
| 3. C | 13. D | 23. D | 33. C | 43. B | 53. B |
| 4. C | 14. C | 24. C | 34. B | 44. C | 54. D |
| 5. D | 15. B | 25. A | 35. D | 45. D | 55. C |
| 6. D | 16. B | 26. A | 36. B | 46. A | 56. C |
| 7. C | 17. A | 27. C | 37. C | 47. C | 57. C |
| 8. D | 18. C | 28. B | 38. B | 48. C | 58. C |
| 9. A | 19. B | 29. A | 39. C | 49. D | 59. B |
| 10. D | 20. B | 30. B | 40. A | 50. C | 60. C |

# Solutions for Exam 1–Morning Section

**1.**
$$\frac{N}{N_0} = \left(\tfrac{1}{2}\right)^{\frac{t}{t_{1/2}}}$$

$$0.01 = \left(\tfrac{1}{2}\right)^{\frac{t}{4.3 \text{ days}}}$$

$$\ln(0.01) = \left(\frac{t}{4.3 \text{ days}}\right)\ln\tfrac{1}{2}$$

$$t = (4.3 \text{ days})\left(\frac{\ln 0.01}{\ln 0.5}\right)$$

$$= 28.6 \text{ days} \quad (29 \text{ days})$$

**Answer is C.**

**2.**
$$(1)(2) - (4)(3) = -10$$

**Answer is A.**

**3.** Solve by factoring.

$$\lim_{x \to 3}\frac{x^2 - 9}{x - 3} = \lim_{x \to 3}\frac{(x+3)(x-3)}{x - 3}$$

$$= \lim_{x \to 3}(x + 3)$$

$$= 6$$

**Answer is C.**

**4.** The magnitude of vector $\mathbf{V}$ is

$$V = \sqrt{(1)^2 + (2)^2 + (1)^2} = \sqrt{6}$$

The $x$-direction cosine is

$$\cos\phi_x = \frac{V_x}{V} = \frac{1}{\sqrt{6}}$$

$$\theta = \cos^{-1}\left(\frac{1}{\sqrt{6}}\right)$$

$$= 65.9° \quad (66°)$$

**Answer is C.**

**5.**
$$\int_2^\infty \frac{1}{x^2}\,dx = -\frac{1}{x}\bigg|_2^\infty = \frac{-1}{\infty} - \frac{-1}{2}$$

$$= 1/2$$

**Answer is C.**

**6.** The magnitudes of the two vectors are

$$V_1 = \sqrt{(1)^2 + (2)^2 + (1)^2} = \sqrt{6}$$

$$V_2 = \sqrt{(1)^2 + (3)^2 + (-7)^2} = \sqrt{59}$$

$$\phi = \cos^{-1}\left(\frac{(1)(1) + (2)(3) + (1)(-7)}{\left(\sqrt{6}\right)\left(\sqrt{59}\right)}\right) = 90°$$

**Answer is C.**

**7.**
$$A = \int_0^1 e^x\,dx = e^x\bigg|_0^1 = e^1 - e^0$$

$$\approx 1.718 \quad (1.7)$$

**Answer is B.**

**8.** The first and second derivatives are

$$y' = x^3 - 3x + 2$$
$$y'' = 3x^2 - 3$$

For a critical point, $y' = 0$. By inspection (based on the four answer choices), $y' = 0$ at $x = 1$ and $x = -2$.

$$y''(1) = (3)(1)^2 - 3 = 0$$
$$y''(-2) = (3)(-2)^2 - 3 = 9$$
$$y(-2) = \left(\tfrac{1}{4}\right)(-2)^4 - (1.5)(-2)^2 + (2)(-2) + 5 = -1$$

**Answer is A.**

**9.**
$$\theta = \tan^{-1}\left(\tfrac{1}{2}\right) - \tan^{-1}\left(-\tfrac{2}{3}\right) = 60.26°$$

**Answer is C.**

**10.**
$$\left(y^2 + y + \tfrac{1}{4}\right) + (x^2 - 2x + 1) = 5 + \tfrac{1}{4} + 1$$
$$= 25/4$$
$$\left(y + \tfrac{1}{2}\right)^2 + (x - 1)^2 = 25/4$$

This is a circle centered at $(1, -\!1/2)$, so $y_{\max}$ is at $x = 1$.

**Answer is C.**

**11.**

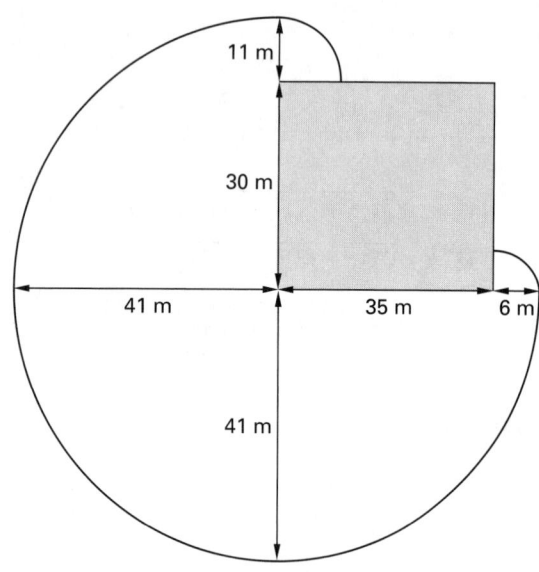

$$\tfrac{3}{4}\pi(41 \text{ m})^2 + \tfrac{1}{4}\pi(11 \text{ m})^2 + \tfrac{1}{4}\pi(6 \text{ m})^2$$
$$= 4084.1 \text{ m}^2 \quad (4080 \text{ m}^2)$$

**Answer is D.**

**12.** Differentiating,

$$y' = 27x^2 + 2x - 15$$
$$y'' = 54x + 2 = 0$$
$$x = -\frac{2}{54} = -0.037$$
$$y(-0.037) = 32.56$$
$$(-0.037, 32.56)$$

**Answer is C.**

**13.** $|\mathbf{R}| = \sqrt{(1+2-1)^2 + (3+7+4)^2 + (4-1+2)^2}$
$$= \sqrt{4 + 196 + 25}$$
$$= \sqrt{225}$$
$$= 15$$

**Answer is C.**

**14.** Multiplying through by 2 gives

$$x'' + 8x' + 16x = 10$$

The characteristic equation is

$$r^2 + 8r + 16 = 0$$

The roots of the characteristic equation are

$$r_1 = r_2 = -4$$

The homogeneous (natural) response is

$$x_{\text{natural}} = Ae^{-4t} + Bte^{-4t}$$

By inspection, $x = {}^5/_8$ is a particular solution that solves the nonhomogeneous equation, so the total response is

$$x = Ae^{-4t} + Bte^{-4t} + \tfrac{5}{8}$$

Since $x = 1$ at $t = 0$,

$$1 = Ae^0 + \tfrac{5}{8}$$
$$A = \tfrac{3}{8}$$

Differentiating $x$,

$$x' = \left(\tfrac{3}{8}\right)(-4)e^{-4t} + B\left(-4te^{-4t} + e^{-4t}\right) + 0$$

Since $x' = 0$ at $t = 0$,

$$0 = -\tfrac{3}{2} + B(0 + 1)$$
$$B = \tfrac{3}{2}$$
$$x = \tfrac{3}{8}e^{-4t} + \tfrac{3}{2}te^{-4t} + \tfrac{5}{8}$$

**Answer is D.**

**15.** Since $y'' = 14$, this is a minimum value at this point.

$$y' = 14x - 3 = 0$$
$$x = 3/14$$
$$y\left(\frac{3}{14}\right) = (7)\left(\frac{3}{14}\right)^2 - (3)\left(\frac{3}{14}\right) + 8$$
$$= (7)\left(\frac{9}{196}\right) - \frac{9}{14} + 8$$
$$= 215/28$$

**Answer is D.**

**16.** From the double angle formulas,

$$\sin 2\theta = 2 \sin \theta \cos \theta$$

**Answer is A.**

**17.**

Use Euler's formula.

$$e^{i\theta} = \cos\theta + i\sin\theta$$

$$\theta = \pi/2$$

$$e^{i\frac{\pi}{2}} = i$$

$$i^i = \left(e^{i\frac{\pi}{2}}\right)^i = e^{-\frac{\pi}{2}}$$

**Answer is D.**

**18.** A graph of the area is shown.

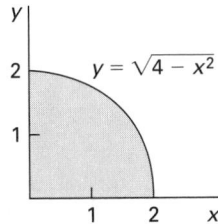

Since $y = \sqrt{4 - x^2}$ is the equation for the top half of a circle, the volume after a revolution will be a hemisphere.

$$V = \tfrac{1}{2}V_{\text{sphere}} = \left(\tfrac{1}{2}\right)\left(\tfrac{4}{3}\right)\pi r^3$$

$$= \left(\tfrac{1}{2}\right)\left(\tfrac{4}{3}\right)\pi(2)^3$$

$$= 16.76 \quad (17)$$

**Answer is C.**

**19.** $\quad p\{\text{at least } 2\} = p\{2 \text{ or } 3 \text{ or } 4 \text{ or } 5\}$

$$= 1 - p\{0 \text{ or } 1\}$$

$$= 1 - p\{0\} - p\{1\}$$

Use binomial distribution to find the individual probabilities.

$$p = \text{probability of heads} = 0.5$$

$$q = \text{probability of tails} = 0.5$$

$$p\{0\} = (0.5)^5 = 0.03125$$

$$p\{1\} = \binom{5}{1}(0.5)^4(0.5)^1 = 0.15625$$

$$p\{\text{at least } 2\} = 1 - 0.03125 - 0.15625$$

$$= 0.8125 \quad (0.81)$$

**Answer is C.**

**20.** $\quad\quad \text{mean} = \dfrac{1 + 4 + 7}{3} = 4$

The standard deviation is

$$\sigma = \sqrt{\dfrac{(1-4)^2 + (4-4)^2 + (7-4)^2}{3}} = \sqrt{6}$$

$$= 2.45 \quad (2.5)$$

Notice that $n - 1$ would have been used if the sample standard deviation had been requested.

**Answer is A.**

**21.** $\quad \text{probability} = \dfrac{\text{number of orange balls}}{\text{total number of balls}} = \dfrac{7}{17}$

$$= 0.4118 \quad (0.41)$$

**Answer is D.**

**22.** First calculate the probability of rolling no sixes. Then subtract that value from 1 to find the probability of rolling at least one six.

$$p\{\text{five dice, no sixes}\} = p\{\text{one die, no six}\}^5$$

$$= \left(\dfrac{5}{6}\right)^5$$

$$= 0.402$$

$$p\left\{\begin{array}{c}\text{five dice, at}\\\text{least one six}\end{array}\right\} = 1 - p\{\text{five dice, no sixes}\}$$

$$= 1 - 0.402$$

$$= 0.598 \quad (0.60)$$

**Answer is C.**

**23.** The mean is

$$\overline{x} = \dfrac{\sum x_i}{n} = \dfrac{2 + 7 + 9 + 12 + 34}{5} = 12.8$$

The sample standard deviation is

$$\sigma = \sqrt{\dfrac{\sum (x_i - \overline{x})^2}{n - 1}}$$

$$= \sqrt{\dfrac{\begin{array}{c}(2 - 12.8)^2 + (7 - 12.8)^2 + (9 - 12.8)^2\\+ (12 - 12.8)^2 + (34 - 12.8)^2\end{array}}{5 - 1}}$$

$$= 12.397 \quad (13)$$

**Answer is B.**

**24.** The mean of the Poisson distribution, $\lambda$, is 5 jams. The probability of 3 jams is

$$p\{3\} = \dfrac{e^{-\lambda}\lambda^x}{x!} = \dfrac{e^{-5}(5)^3}{3!} = 0.140$$

**Answer is B.**

**25.**  First calculate the following values.

$$\sum x_i = 2 + 4 + 1 + 5 = 12$$

$$\sum y_i = 2 + 7 + 11 + 9 = 29$$

$$\sum x_i^2 = (2)^2 + (4)^2 + (1)^2 + (5)^2 = 46$$

$$\sum x_i y_i = (2)(2) + (4)(7) + (1)(11) + (5)(9) = 88$$

The slope is

$$b = \frac{\sum x_i y_i - \dfrac{(\sum x_i)(\sum y_i)}{n}}{\sum x_i^2 - \dfrac{(\sum x_i)^2}{n}}$$

$$= \frac{88 - \dfrac{(12)(29)}{4}}{46 - \dfrac{(12)^2}{4}}$$

$$= 0.1 \quad (0)$$

**Answer is C.**

**26.**  Curve fitting is the method used to represent empirical data using a model based on mathematical equations. With the correct model and calculus, one can determine important characteristics of the data, such as the rate of change anywhere on the curve (first derivative), the local minimum and maximum points of the function (zeros of the first derivative), and the area under the curve (integral).

**Answer is B.**

**27.  Answer is D.**

**28.**

| before | | after | |
|---|---|---|---|
| Zn | 0 | Zn | +2 |
| H | +1 | H | 0 |
| S | +6 | S | +6 |
| O | −2 | O | −2 |

Hydrogen is reduced, and zinc is oxidized.

**Answer is C.**

**29.**  Three moles of oxidizing agent $H^+$ are produced per mole of $H_3PO_4$. One mole of reducing agent $OH^-$ is produced per mole of NaOH. Three moles will be neutralized.

**Answer is D.**

**30.**  Electrons in the external circuit flow from the anode to the cathode, so the anode, as a source of electrons, must be where oxidation (a loss of electrons) occurs.

**Answer is B.**

**31.**  Beryllium, magnesium, and calcium are members of alkaline earth metals (group IIA on the periodic table).

**Answer is D.**

**32.**  Chlorine has a value of −1; therefore, Ca has a valence of +2.

**Answer is D.**

**33.**  $APbO_2 + BH_2SO_4 + CPb \rightleftharpoons DH_2O + EPbSO_4$

$$O: 2A + 4B = D + 4E$$

$$Pb: A + C = E$$

$$H: 2B = 2D; \ B = D$$

$$S: B = E; \ B = D = E$$

$$2A + 4D = D + 4D$$

$$A = \frac{D}{2}$$

$$\frac{D}{2} + C = D$$

$$C = \frac{D}{2}$$

$$A = C = 1$$

$$D = E = B = 2$$

$$PbO_2 + 2H_2SO_4 + Pb \rightleftharpoons 2H_2O + 2PbSO_4$$

**Answer is D.**

**34.**  Consider 100 g of the compound.

$$n_O = \frac{13.7 \text{ g}}{16 \dfrac{\text{g}}{\text{mol}}} = 0.8563 \text{ mol}$$

$$n_C = \frac{20.5 \text{ g}}{12 \dfrac{\text{g}}{\text{mol}}} = 1.71 \text{ mol}$$

$$n_H = \frac{5.1 \text{ g}}{1 \dfrac{\text{g}}{\text{mol}}} = 5.1 \text{ mol}$$

$$n_{Cl} = \frac{60.7 \text{ g}}{35.45 \dfrac{\text{g}}{\text{mol}}} = 1.71 \text{ mol}$$

$n_O$ is the lowest common denominator.

$$\frac{n_O}{n_O} = \frac{0.8563}{0.8563} = 1$$

$$\frac{n_C}{n_O} = \frac{1.71}{0.8563} = 1.997 \approx 2$$

$$\frac{n_H}{n_O} = \frac{5.1}{0.8563} = 5.956 \approx 6$$

$$\frac{n_{Cl}}{n_O} = \frac{1.71}{0.8563} = 1.997 \approx 2$$

$$C_2H_6OCl_2$$

**Answer is C.**

**35.** Each compound ionizes to form 1 mol of each ion.

$$NaOH + HCl \longrightarrow NaCl + H_2O$$
$$Na^+ + OH^- + H^+ + Cl^- \longrightarrow Na^+ + Cl^- + H_2O$$

**Answer is C.**

**36.** C-12 has a mass of 12 g/mol.

$$\frac{mass}{molecule} = \frac{mass\ per\ mol}{molecules\ per\ mol}$$

$$= \left(12\ \frac{g}{mol}\right)\left(\frac{1\ mol}{6.02 \times 10^{-23}\ molecules}\right)$$

$$= 1.99 \times 10^{-23}\ g \quad (2.0 \times 10^{-23}\ g)$$

**Answer is B.**

**37.** $2s^2 + 14s - 36 = (2s - 4)(s + 9) = 0$ [at pole]
$$\rightarrow \text{poles at } s = -9, 2$$
$$s = 2\ Hz$$

Positive poles are unstable.

**Answer is B.**

**38.**

Compare the system with the general system sketched.

$$G_R(s) = 1$$
$$G_C(s) = 15s$$
$$G_1(s) = 21/s$$
$$L(s) = 0$$
$$G_2(s) = 1$$
$$H(s) = \frac{s^2}{35} + \frac{3s}{35} + \frac{1}{4}$$
$$\frac{C(s)}{R(s)} = \frac{G_C(s)G_1(s)G_2(s)G_R(s)}{1 + G_C(s)G_1(s)G_2(s)H(s)}$$

$$= \frac{(5s)\left(\frac{7}{s}\right)(1)(1)}{1 + (5s)\left(\frac{7}{s}\right)(1)\left(\frac{s^2}{35} + \frac{3s}{35} + \frac{1}{4}\right)}$$

$$= \frac{35}{s^2 + 3s + \frac{39}{4}} \qquad \text{I}$$

For a second-order control system,

$$\frac{C(s)}{R(s)} = \frac{kw_n^2}{s^2 + 2\zeta w_n s + w_n^2} \qquad \text{II}$$

$k$ is the steady-state gain, $\zeta$ is the damping ratio, and $w_n$ is the natural frequency.

Comparing Eqs. I and II,

$$w_n^2 = 39/4$$

$$w_n = \sqrt{\frac{39}{4}}$$

$$= 3.122 \quad (3.12)$$

**Answer is B.**

**39.** Compare Eqs. I and II from Sol. 38.

$$kw_n^2 = 35$$

$$k = \frac{35}{w_n^2} = \frac{35}{(5)^2} = 1.4$$

**Answer is A.**

**40.** Compare Eqs. I and II from Sol. 38.

$$2\zeta w_n = 3$$

$$\zeta = \frac{3}{2w_n} = \frac{3}{(2)(5)} = 0.3$$

**Answer is B.**

**41.**   $\omega_d = \omega_n \sqrt{1 - \zeta^2} = (5)\sqrt{1 - (0.5)^2} = 4.33$

**Answer is D.**

**42.**   $\omega_p = \omega_n \sqrt{1 - 2\zeta^2} = (5)\sqrt{1 - (2)(0.5)^2} = 3.54$

**Answer is B.**

**43.**   The characteristic equation is the denominator of Eq. I from Sol. 38, and the poles of the system lie at the roots of this equation.

$$s^2 + 3s + \frac{39}{4} = 0$$

$$s = \frac{-3 \pm \sqrt{9 - 39}}{2} = \frac{-3}{2} \pm \frac{\sqrt{30}}{2}\, i$$

Since the real part of both roots is negative, the system is stable. Since the imaginary part is nonzero, the output will oscillate. Oscillations are characteristic of an underdamped system.

**Answer is A.**

**44.**   A baud is a unit of data transmission speed roughly equal to one bit per second. The measure was named after the French engineer J.M.E. Baudot and was initially used to measure the speed of telegraph transmissions. Bauds were used to measure modem speed until the early 1990s. At the lower model speeds available then, the baud rate generally equaled the rate of transmission in bits per second (bps). Higher modem speeds are generally measured in kilobits per second (Kbps).

**Answer is D.**

**45.**   The code is an example of a function call. A function call is similar to a subroutine except that a function always returns a value to the calling program. A function is usually called implicitly by embedding the function call into another statement in place of the returned value rather than having a separate call statement.

**Answer is B.**

**46.**   Professionals may provide services even if conflicts of interest are involved as long as they disclose the conflicts of interest and any influences and biases such conflicts may cause.

**Answer is C.**

**47.**   It is ethical to receive compensation from more than one party if all parties agree to the arrangement.

**Answer is D.**

**48.**   Political contributions can be made if not intended to influence awards or achieve favor.

**Answer is B.**

**49.**   While (A), (C), and (D) are always ethics violations, an engineer is responsible to publicize confidential information about a product or process if not doing so could jeopardize the welfare of society.

**Answer is B.**

**50.**   Although ethical obligations to an employer are important, providing a safe, quality product to the consumer and client that will not jeopardize the welfare of society comes before any responsibilities to an employer.

**Answer is A.**

**51.**   While employers, societies, and all engineers should work to ensure ethical behavior, official regulation is provided by state agencies.

**Answer is D.**

**52.**   Whistle blowing is the act of telling news agencies or regulating boards about flaws in a design or procedure that could jeopardize the public welfare. It is the ethical responsibility of all engineers, although it is not career-enhancing.

**Answer is A.**

**53.**   Compensation is not a reason for ethical behavior.

**Answer is B.**

**54.**   Clients are not required to seek competitive bids. In fact, many engineering societies discourage the use of bidding to procure design services because it is believed that competitive bidding results in lower-quality construction.

**Answer is B.**

**55.**
$$\begin{aligned} i_{\text{annual}} &= (1 + i_{\text{monthly}})^{12} - 1 \\ &= (1 + 0.012)^{12} - 1 \\ &= 0.1538 \quad (15\%) \end{aligned}$$

**Answer is D.**

**56.** With simple interest,

$$F = P(1 + i)$$
$$= (\$300)\big(1 + (0.12)(3)\big) = \$408$$

**Answer is B.**

**57.**
$$P = P_0(1 + i)^n$$
$$2P_0 = P_0(1 + i)^n$$
$$\log 2 = n \log (1 + i)$$
$$n = \frac{\log 2}{\log (1 + i)} = \frac{\log 2}{\log 1.05}$$
$$= 14.2 \text{ yr} \quad (14 \text{ yr})$$

**Answer is D.**

**58.**
$$P = (\text{net cash flow})(P/A, 6\%, n)$$
$$\$40,000 = (\$10,000 - \$2000)(P/A, 6\%, n)$$
$$(P/A, 6\%, n) = 5$$

Interpolating from the 6% economic factor table, $n = 6.1$ yr.

**Answer is B.**

**59.** A 10-yr period with interest compounded semi-annually is 20 periods.

$$\phi = \frac{r}{k} = \frac{8\%}{2} = 4\%$$
$$P = F(P/F, 4\%, 20) = (\$10,000)(0.4564)$$
$$= \$4564 \quad (\$4560)$$

**Answer is B.**

**60.** The present sum, $P$, is \$500. The interest rate, $i$, is 6%, and in 3 yr there will be three interest periods, $n$. Compute the future sum.

$$F = P(F/P, i\%, n) = P(1 + i)^n$$
$$= (\$500)(1 + 0.06)^3$$
$$= \$595.51 \quad (\$600)$$

**Answer is B.**

**61.** Use the uniform series compound formula.

$$F = A(F/A, i\%, n) = A\left(\frac{(1 + i)^n - 1}{i}\right)$$
$$= (\$500)\left(\frac{(1 + 0.05)^5 - 1}{0.05}\right)$$
$$= \$2762.82 \quad (\$2760)$$

**Answer is C.**

**62.** Use the uniform series capital recovery factor to solve for disbursements.

$$A = P(A/P, i\%, n) = P\left(\frac{i(1 + i)^n}{(1 + i)^n - 1}\right)$$
$$= (\$5000)\left(\frac{(0.08)(1 + 0.08)^5}{(1 + 0.08)^5 - 1}\right)$$
$$= \$1252 \quad (\$1250)$$

**Answer is D.**

**63.** Use the single payment present worth formula.

$$P = F(P/F, i\%, n) = F\frac{1}{(1 + i)^n}$$
$$= (\$800)\left(\frac{1}{(1 + 0.05)^4}\right)$$
$$= \$658.16 \quad (\$660)$$

**Answer is C.**

**64.** The distributed load is replaced by

$$F = \left(\tfrac{1}{2}\right)\left(100 \ \frac{\text{N}}{\text{m}}\right)(4 \text{ m}) = 200 \text{ N}$$

The force is applied at $(^2/_3)(4 \text{ m}) = 8/3$ m from point A.

$$\sum M_B = 0$$
$$\left(10 \text{ m} - \frac{8}{3} \text{ m}\right)(200 \text{ N}) - (10 \text{ m})R_A = 0$$
$$R_A = 146.67 \text{ N} \quad (150 \text{ N})$$

**Answer is D.**

**65.** First, find the reaction.

$$R_{\text{left},y} = R_{\text{right},y}$$
$$= \frac{(7)(35 \text{ N}) + (7)(50 \text{ N}) + (2)(25 \text{ N})}{2}$$
$$= 322.5 \text{ N}$$

Cut the truss vertically through the second section from the left.

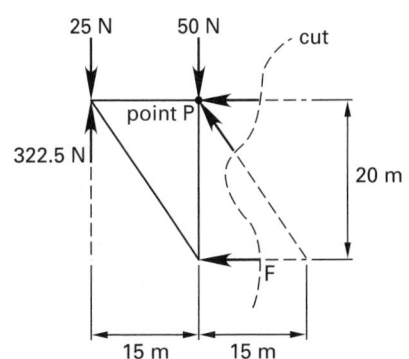

$$\sum M_P = 0$$
$$-(20 \text{ m})F + (15 \text{ m})(25 \text{ N} - 322.5 \text{ N}) = 0$$
$$F = -223.125 \text{ N} \quad (223 \text{ N})$$

**Answer is D.**

**66.** The vertical component in BC is the same as the reaction at B.

$$\sum M_A = 0$$
$$(7 \text{ m})(R_{B,y}) - (3 \text{ m})(2300 \text{ N}) = 0$$
$$R_{B,y} = |BC|_y = 985.7 \text{ N} \quad (990 \text{ N})$$

**Answer is A.**

**67.** $N = W \cos\phi = (784.8 \text{ N})(\cos 40°) = 601.2 \text{ N}$
$$F_{f,s} = N\mu_s = (601.2 \text{ N})(0.2) = 120.2 \text{ N}$$
$$F_g = (784.8 \text{ N})(\sin 40°) = 504.5 \text{ N}$$
$$F_g > F_{f,s}$$

The block is sliding.

$$F_{f,d} = N\mu_d = (601.2 \text{ N})(0.15) = 90.17 \text{ N}$$

**Answer is C.**

**68.** The freebody diagram of the top block is

$$F_{f,5} = \mu N = \mu W = \mu mg$$
$$= (0.15)(5 \text{ kg})\left(9.81 \frac{\text{m}}{\text{s}^2}\right)$$
$$= 7.36 \text{ N}$$

The freebody diagram of the lower block is

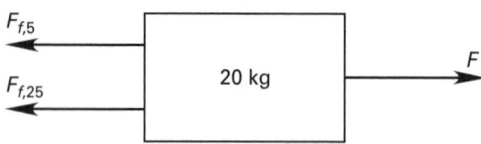

$$F = F_{f,5} + F_{f,25}$$
$$= 7.36 \text{ N} + (0.30)(20 \text{ kg} + 5 \text{ kg})\left(9.81 \frac{\text{m}}{\text{s}^2}\right)$$
$$= 80.9 \text{ N}$$

**Answer is D.**

**69.** Draw the freebody of the lower pulley and spreader bar. All forces are the same because the pulleys are frictionless.

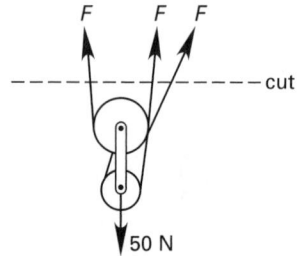

$$F = \frac{50 \text{ N}}{3} = 16.67 \text{ N} \quad (17 \text{ N})$$

**Answer is C.**

**70.** When the block begins to tip, it will tip about the lower right corner. Since the horizontal component of the reaction at this corner is unknown, moments must be taken about the corner to determine $x$.

$$\sum M_{\text{lower right corner}} = 0$$
$$-x(20 \text{ N}) + \left(\frac{4 \text{ m}}{2}\right)(50 \text{ N}) = 0$$
$$x = 5.0 \text{ m}$$

**Answer is D.**

**71.** Impact force is normal to the wall in the $x$ direction only.

$$\frac{v_{x,f}}{v_{x,i}} = e$$
$$v_{x,f} = (0.8)\left(50 \frac{\text{m}}{\text{s}}\right)\left(\frac{3}{\sqrt{(3)^2 + (4)^2}}\right)$$
$$= 24 \text{ m/s}$$

**Answer is A.**

**72.** $\quad s = \int_{t_1}^{t_2} v \, dt = \int_{0.2}^{0.3}\left(12t^4 + \frac{7}{t}\right)dt$
$$= \left(\frac{12}{5}t^5 + 7\ln t\right)_{0.2}^{0.3}$$
$$= \left(\frac{12}{5}\right)\left((0.3)^5 - (0.2)^5\right) + 7\ln\frac{3}{2}$$
$$= 2.84 \quad (2.8)$$

**Answer is C.**

**73.**
$$E_{\text{spring}} = E_{\text{projectile}}$$

$$\tfrac{1}{2}k(\Delta x)^2 = \frac{m\text{v}^2}{2}$$

$$\text{v} = \sqrt{\frac{k}{m}}\,\Delta x$$

$$= \sqrt{\frac{\left(10\ \dfrac{\text{N}}{\text{cm}}\right)\left(100\ \dfrac{\text{cm}}{\text{m}}\right)}{1\ \text{kg}}}$$

$$\times (5\ \text{cm})\left(\frac{1\ \text{m}}{100\ \text{cm}}\right)$$

$$= 1.58\ \text{m/s} \quad (1.6\ \text{m/s})$$

**Answer is B.**

**74.** $P = \dfrac{\Delta E_k}{\Delta t} = \dfrac{\tfrac{1}{2}m\left(\text{v}_1^2 - \text{v}_2^2\right)}{\Delta t}$

$$= \frac{\left(\tfrac{1}{2}\right)(213\,000\ \text{kg})\left(\left(9020\ \dfrac{\text{m}}{\text{s}}\right)^2 - \left(5100\ \dfrac{\text{m}}{\text{s}}\right)^2\right)}{48\ \text{s}}$$

$$= 122.8 \times 10^9\ \text{W} \quad (120\ \text{GW})$$

**Answer is B.**

**75.** The initial kinetic energy is

$$E_k = \frac{m\text{v}^2}{2}$$

$$= \left(\frac{(32\ \text{kg})\left(3600\ \dfrac{\text{m}}{\text{s}}\right)^2}{2}\right)\left(\frac{1\ \text{MJ}}{10^6\ \dfrac{\text{kg}\cdot\text{m}^2}{\text{s}^2}}\right)$$

$$= 207\ \text{MJ} \quad (210\ \text{MJ})$$

After launch, some of the kinetic energy will be transformed into potential energy, but the total energy will be unchanged.

**Answer is C.**

**76.** $\epsilon_{\text{th}} = \alpha\Delta T = \left(6 \times 10^{-6}\ \dfrac{\text{cm}}{\text{cm}\cdot{}^\circ\text{C}}\right)(60{}^\circ\text{C})$

$$= 0.00036$$

$$\sigma_{\text{th}} = E\epsilon_{\text{th}} = \left(30 \times 10^6\ \frac{\text{N}}{\text{cm}^2}\right)(0.00036)$$

$$= 10\,800\ \text{N/cm}^2 \quad (11\,000\ \text{N/cm}^2)$$

**Answer is B.**

**77.**
$$\delta = \epsilon_{\text{th}}L_o = (0.00036)(30\ \text{cm})$$

$$= 0.0108\ \text{cm} \quad (0.1\ \text{cm})$$

**Answer is C.**

**78. Answer is C.**

**79.** The use of variable $c$ to represent the entire beam depth is nonstandard.

$$\sigma_{x,\text{max}} = \frac{M\left(\dfrac{c}{2}\right)}{I} = Pxc/2I$$

**Answer is A.**

**80.** The cable length is

$$L = \sqrt{(3\ \text{m})^2 + (4\ \text{m})^2} = 5\ \text{m}$$

The cable tension is found by similar triangles.

$$T = \left(\frac{5\ \text{m}}{3\ \text{m}}\right)(1000\ \text{N}) = 1667\ \text{N}$$

The elongation is

$$\delta = \frac{TL}{AE} = \frac{(1667\ \text{N})(5\ \text{m})\left(100\ \dfrac{\text{cm}}{\text{m}}\right)}{(2\ \text{cm}^2)\left(1.5 \times 10^6\ \dfrac{\text{N}}{\text{cm}^2}\right)}$$

$$= 0.278\ \text{cm} \quad (0.28\ \text{cm})$$

**Answer is C.**

**81.** A high slenderness ratio means that the beam will fail due to bending moment. The symmetry of the beam indicates that reactions will be the same at either end. $M_{\text{max}}$ is at the center, where the slope of the deflection curve is zero.

**Answer is D.**

**82.** For this material, the stress required to continue the deformation drops markedly at yield.

$$\sigma_y = 31\ \text{kN/cm}^2$$

**Answer is D.**

**83.** The elastic limit is very close to the yield point.

$$\sigma_{\text{elastic limit}} = 31\ \text{kN/cm}^2$$

**Answer is C.**

**84.**
$$A = \pi r^2 = \frac{\pi d^2}{4}$$

$$= \frac{\pi \left( (30 \text{ mm}) \left( \frac{m}{1000 \text{ mm}} \right) \right)^2}{4}$$

$$= 7.07 \times 10^{-4} \text{ m}^2$$

$$\sigma = \frac{P}{A} = \frac{150 \text{ kN} \left( \frac{1000 \text{ N}}{\text{kN}} \right)}{7.07 \times 10^{-4} \text{ m}^2}$$

$$= 212.2 \text{ Pa}$$

$$\epsilon = \frac{\delta}{L} = \frac{1.0 \text{ mm}}{200 \text{ mm}}$$

$$= 0.005 \text{ mm/mm}$$

$$E_{\text{alum}} = \frac{\sigma}{\epsilon} = \frac{212.2 \times 10^6 \text{ Pa}}{0.005 \frac{\text{mm}}{\text{mm}}}$$

$$= 42\,440 \times 10^6 \text{ Pa} \quad (42.4 \text{ GPa})$$

**Answer is C.**

**85.** All metals have similar stress-strain curves, so deformation into the plastic region will produce residual strain in all metals.

**Answer is A.**

**86.** An impact test measures the energy needed to fracture the test sample. This is a toughness parameter.

**Answer is C.**

**87.** Cast iron is iron with 2% or more carbon. Steels have from 0.01% to 2% carbon.

**Answer is A.**

**88.** Excess water decreases concrete strength.

**Answer is A.**

**89.**
$$\lambda = \frac{2\pi r}{n} = \frac{(2\pi)(0.75\text{Å})}{4}$$
$$= 1.18\text{Å}$$

**Answer is B.**

**90.** An orbital is a spatial approximation of a probability function describing the likelihood of finding an electron in a certain region. It may be occupied by 0, 1, or 2 electrons with opposing spin directions.

**Answer is D.**

**91.** For corrosion to occur, there must be an anode and a cathode connected through an electrolyte.

**Answer is D.**

**92.** Since solid and liquid phases are present simultaneously, the number of phases, $P$, is 2. The only element involved is water, so the number of compounds, $C$, is 1.
$$P + F = C + 2$$
$$2 + F = 1 + 2$$
$$F = 1$$

**Answer is C.**

**93.** Use of a venturi meter implies pipe flow, which means
$$(N_{\text{Re}})_{\text{actual}} = (N_{\text{Re}})_{\text{model}}$$
The units given for the viscosity are for absolute viscosity, not kinematic viscosity.

$$\frac{\rho_{\text{air}} \text{v}_{\text{actual}} L_{\text{actual}}}{\mu_{\text{air}}} = \frac{\rho_{\text{water}} \text{v}_{\text{model}} L_{\text{model}}}{\mu_{\text{water}}}$$

$$\frac{\text{v}_{\text{model}}}{\text{v}_{\text{actual}}} = \left( \frac{\mu_{\text{water}}}{\mu_{\text{air}}} \right) \left( \frac{\rho_{\text{air}}}{\rho_{\text{water}}} \right) \left( \frac{L_{\text{actual}}}{L_{\text{model}}} \right)$$

$$= \left( \frac{9.82 \times 10^{-4} \frac{\text{N·s}}{\text{m}^2}}{1.82 \times 10^{-5} \frac{\text{N·s}}{\text{m}^2}} \right) \left( \frac{1.20 \frac{\text{kg}}{\text{m}^3}}{1000 \frac{\text{kg}}{\text{m}^3}} \right) \left( \frac{5}{1} \right)$$

$$= 0.3237 \quad (0.32)$$

**Answer is A.**

**94.** Assume partial similarity based on the Froude number since the fluid is not mentioned. (Froude number similarity is appropriate for modeling of surface vessels.)

$$\text{Fr} = \frac{\text{v}^2}{Lg}$$

$$\frac{\text{v}_m^2}{L_m g} = \frac{\text{v}^2}{Lg}$$

$$\text{v}_m = \text{v}\sqrt{\frac{L_m}{L}} = \left[ \frac{\left( 40 \frac{\text{km}}{\text{h}} \right) \left( 1000 \frac{\text{m}}{\text{km}} \right)}{3600 \frac{\text{s}}{\text{h}}} \right] \left( \sqrt{\frac{1}{50}} \right)$$

$$= 1.57 \text{ m/s} \quad (1.6 \text{ m/s})$$

**Answer is C.**

**95.**

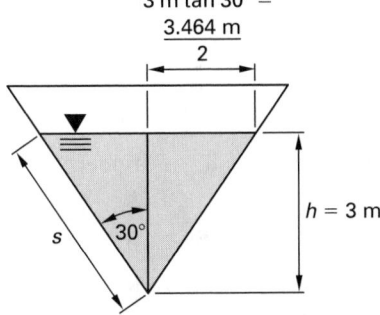

$$s = \frac{3 \text{ m}}{\cos 30^\circ} = 3.464 \text{ m}$$

$$\begin{array}{c} \text{area in flow} \\ \text{at full capacity} \end{array} = \left(\tfrac{1}{2}\right)(3.464 \text{ m})(3 \text{ m}) = 5.196 \text{ m}^2$$

$$r_h = \frac{\text{area in flow}}{\text{wetted perimeter}} = \frac{5.196 \text{ m}^2}{(2)(3.464 \text{ m})}$$

$$= 0.75 \text{ m}$$

**Answer is D.**

**96.**  $\rho_{\text{water}} L = \rho_{\text{rod}}(24 \text{ cm})$

$$L = \left(\frac{\rho_{\text{rod}}}{\rho_{\text{water}}}\right)(24 \text{ cm}) = (\text{SG})(24 \text{ cm})$$

$$= (0.6)(24 \text{ cm})$$

$$= 14.4 \text{ cm} \quad (14 \text{ cm})$$

**Answer is B.**

**97.**  $F = \overline{p}A = \rho g \overline{h} A$

$$= \rho g \left(\tfrac{1}{2}\right)(8 \text{ m} + 10 \text{ m})((5 \text{ m})(2 \text{ m}))$$

$$= \left(1000 \frac{\text{kg}}{\text{m}^3}\right)\left(9.81 \frac{\text{m}}{\text{s}^2}\right)(90 \text{ m}^3)$$

$$= 883\,000 \text{ N} \quad (880 \text{ kN})$$

**Answer is C.**

**98.** Cohesive forces dominate in mercury. This depresses the mercury level in the tube.

**Answer is B.**

**99.**  $(\text{SG})_{\text{fluid}} = (0.4)(0.8) + (0.6)(1.0) = 0.92$

$$\Delta p = \rho g \Delta h$$

$$\Delta h = \frac{\Delta p}{\rho g} = \frac{\Delta p}{(\text{SG})_{\text{fluid}} \gamma_{\text{H}_2\text{O}} g}$$

$$= \frac{42\,700 \text{ Pa}}{(0.92)\left(1000 \frac{\text{kg}}{\text{m}^3}\right)\left(9.81 \frac{\text{m}}{\text{s}^2}\right)}$$

$$= 4.73 \text{ m} \quad (470 \text{ cm})$$

**Answer is C.**

**100.** Determine the specific gas constant.

$$R = \frac{R^*}{\text{MW}} = \frac{8314 \dfrac{\text{J}}{\text{kmol·K}}}{44 \dfrac{\text{kg}}{\text{kmol}}}$$

$$= 189.0 \text{ J/kg·K}$$

Solve for the density.

$$\rho = \frac{p}{RT}$$

$$= \frac{1.38 \times 10^5 \text{ Pa}}{\left(189.0 \dfrac{\text{J}}{\text{kg·K}}\right)(66^\circ\text{C} + 273^\circ)}$$

$$= 2.15 \text{ kg/m}^3$$

Solve for specific gravity as a function of the density of the carbon dioxide and the density of STP air.

$$\text{SG} = \frac{2.15 \dfrac{\text{kg}}{\text{m}^3}}{1.29 \dfrac{\text{kg}}{\text{m}^3}} = 1.67 \quad (1.7)$$

**Answer is C.**

**101.**

$$p_a = p_v + \rho g h$$

$$= 2.34 \text{ kPa} + \frac{\left(998 \dfrac{\text{kg}}{\text{m}^3}\right)\left(9.81 \dfrac{\text{m}}{\text{s}^2}\right)(10.5 \text{ m})}{1000 \dfrac{\text{Pa}}{\text{kPa}}}$$

$$= 105.1 \text{ kPa} \quad (100 \text{ kPa})$$

**Answer is A.**

**102.**  $\quad i = \dfrac{\Delta Q}{\Delta t} = \dfrac{6 \text{ C}}{2 \text{ s}} = 3 \text{ A}$

**Answer is B.**

**103.**  $\quad a = \dfrac{N_p}{N_s} = \dfrac{200 \text{ turns}}{20 \text{ turns}} = 10$

$$V_s = \frac{V_p}{a} = \frac{120 \text{ V}}{10} = 12 \text{ V}$$

**Answer is B.**

**104.**  $\quad \dfrac{N_p}{N_s} = a = \dfrac{I_s}{I_p} = \dfrac{0.3 \text{ A}}{30 \text{ A}}$

$$= 0.01 \quad (1{:}100)$$

**Answer is A.**

**105.**
$$P_p = I_p^2 R_p$$

$$I_p = \sqrt{\frac{P_p}{R_p}} = \sqrt{\frac{5 \text{ W}}{2000 \ \Omega}} = 0.05 \text{ A}$$

$$I_s = aI_p = (15)(0.05 \text{ A}) = 0.75 \text{ A}$$

$$R_s = \frac{P_s}{I_s^2} = \frac{P_p}{I_s^2} = \frac{5 \text{ W}}{(0.75 \text{ A})^2}$$

$$= 8.89 \ \Omega \quad (8.9 \ \Omega)$$

**Answer is A.**

**106.** discharged energy = energy stored in capacitor

$$E = \int_0^V P \, dt = \int_0^V VI \, dt$$

$$= \int_0^V V \left( C \frac{dV}{dt} \right) dt$$

$$= \tfrac{1}{2} CV^2$$

**Answer is D.**

**107.** The impedance is

$$|\mathbf{Z}| = \sqrt{(4 \ \Omega)^2 + (7 \ \Omega)^2} = 8.06 \ \Omega$$

The impedance angle (voltage leading current) is

$$\phi_Z = \arctan\left(\frac{-7 \ \Omega}{4 \ \Omega}\right) \approx -60°$$

$$\mathbf{I} = \frac{\mathbf{V}}{\mathbf{Z}} = \frac{120\angle 0° \text{ V}}{8.06 \angle -60° \ \Omega} = 14.88 \angle 60° \text{ A}$$

Since $\mathbf{V}$ has a zero phase angle, the current leads the voltage.

**Answer is D.**

**108.**
$$Z = \sqrt{R^2 + (X_L - X_C)^2}$$

$$= \sqrt{\left(\tfrac{1}{4} \ \Omega\right)^2 + \left(\tfrac{1}{6} \ \Omega - \tfrac{1}{3} \ \Omega\right)^2}$$

$$= 0.3 \ \Omega$$

$$\phi_Z = \arctan\left(\frac{X_L - X_C}{R}\right)$$

$$= \arctan\left(\frac{\tfrac{1}{6} \ \Omega - \tfrac{1}{3} \ \Omega}{\tfrac{1}{4} \ \Omega}\right)$$

$$= \arctan(-0.6667)$$

$$= -33.69°$$

$$\mathbf{V} = \mathbf{IZ} = (4\angle -20° \text{ A})(0.3\angle -33.69° \ \Omega)$$

$$= 1.2\angle -53.69° \text{ V} \quad (1.2\angle -53.7° \text{ V})$$

**Answer is B.**

**109.** This is a DC source in series with a capacitor, so the steady-state current is zero.

**Answer is A.**

**110.** This is a simple voltage divider.

$$V_{4 \ \Omega} = (120 \text{ V}) \left(\frac{4 \ \Omega}{8 \ \Omega + 4 \ \Omega}\right) = 40 \text{ V}$$

**Answer is B.**

**111.** At resonance, the reactive impedances cancel and the circuit impedance is equal to the resistance.

**Answer is A.**

**112.** $\Delta U = C_v \Delta T$ by definition for both processes. For an isothermal process, $\Delta T = 0$, so $\Delta U = 0$.

**Answer is D.**

**113.**
$$\eta = 1 - \frac{T_L}{T_H} = 1 - \frac{150°\text{C} + 273}{550°\text{C} + 273}$$

$$= 0.486 \quad (49\%)$$

**Answer is B.**

**114.** $\Delta H = nC_p \Delta T$ by definition.

**Answer is D.**

**115.** Entropy is temperature dependent. It theoretically approaches zero as the temperature approaches absolute zero.

**Answer is C.**

**116.** This situation describes a throttling process. Entropy increases, temperature decreases, and enthalpy is unchanged.

**Answer is A.**

**117.** Answer is A.

**118.**
$$\frac{p_1 V_1}{T_1} = \frac{p_2 V_2}{T_2}$$

$$p_2 = p_1 \left(\frac{V_1}{V_2}\right)\left(\frac{T_2}{T_1}\right)$$

$$= p_1 \left(\frac{V_1}{\tfrac{1}{2}V_1}\right)\left(\frac{2T_1}{T_1}\right)$$

$$= 4p_1$$

**Answer is C.**

**119.** Saturated steam should not be treated as an ideal gas, but the pressure will nevertheless decrease as the volume increases.

**Answer is A.**

**120.** In an adiabatic process, $Q = 0$, but enthalpy and work could still be affected by $\Delta(pV)$. An isentropic process is both reversible and adiabatic.

**Answer is B.**

# Solutions for Exam 1–Afternoon Section

**1.**
$$x^2 + 4x + y^2 + 8y = 0$$
$$x^2 + 4x + 4 + y^2 + 8y + 16 = 4 + 16$$
$$(x + 2)^2 + (y + 4)^2 = 20$$

The center is at $(-2, -4)$.

**Answer is D.**

**2.**
$$r = \sqrt{20} = 2\sqrt{5}$$

**Answer is B.**

**3.** Line $L_{CO}$, through the center of the circle and origin, is perpendicular to $L_{TO}$, the tangent line to the circle through the origin.

$$m_{TO} = \frac{-1}{m_{CO}} = \frac{-1}{\frac{y}{x}} = \frac{-1}{\frac{4}{2}} = -1/2$$

**Answer is C.**

**4.** A derivative appears, so it is a differential equation. The highest derivative taken is the first, so it is a first-order differential equation. No products of derivatives of different variables or powers of a single derivative appear, so it is a linear first-order differential equation.

**Answer is C.**

**5.**
$$2y' = 3xy + 1$$
$$y' - \tfrac{3}{2}xy = 1/2$$
$$u(x) = e^{\int_0^x -\frac{3}{2}x\,dx} = e^{-\frac{3}{4}x^2}$$

**Answer is D.**

**6.**
$$y = \frac{1}{u(x)} \left( \int u(x)g(x)dx + C \right)$$
$$= \left( \frac{1}{e^{-\frac{3}{4}x^2}} \right) \left( \int e^{-\frac{3}{4}x^2} \left(\tfrac{1}{2}\right) dx + C \right)$$
$$= \left( e^{\frac{3}{4}x^2} \right) \left( \tfrac{1}{2} \int e^{-\frac{3}{4}x^2} dx + C \right)$$

The integral cannot be expressed as an elementary function. The Taylor series expansion of $e^{-(3/4)x^2}$ could be integrated (ala an error function) to obtain an approximate solution.

**Answer is D.**

**7.** First calculate the following values.
$$\sum x_i = 3 + 3 + 4 + (-1) = 9$$
$$\sum y_i = -5 + (-2) + 3 + 6 = 2$$
$$\sum x_i^2 = (3)^2 + (3)^2 + (4)^2 + (-1)^2 = 35$$
$$\sum y_i^2 = (-5)^2 + (-2)^2 + (3)^2 + (6)^2 = 74$$
$$\sum x_i y_i = (3)(-5) + (3)(-2) + (4)(3) + (-1)(6)$$
$$= -15$$

The correlation coefficient is given by

$$r = \frac{\sum x_i y_i - \frac{\left(\sum x_i\right)\left(\sum y_i\right)}{n}}{\sqrt{\left(\sum x_i^2 - \frac{\left(\sum x_i\right)^2}{n}\right)\left(\sum y_i^2 - \frac{\left(\sum y_i\right)^2}{n}\right)}}$$

$$= \frac{(-15) - \frac{(9)(2)}{4}}{\sqrt{\left(35 - \frac{(9)^2}{4}\right)\left(74 - \frac{(2)^2}{4}\right)}}$$

$$= -0.594 \quad (-0.59)$$

**Answer is C.**

**8.** This is a binomial distribution problem.
$$p = 0.02$$
$$q = 1 - p = 0.98$$
$$p\{3\} = \left( \frac{n!}{x!(n-x)!} \right) p^x q^{n-x}$$
$$= \left( \frac{20!}{(3!)(20-3)!} \right) (0.02)^3 (0.98)^{20-3}$$
$$= 0.00647 \quad (0.0065)$$

**Answer is D.**

**9.** The number of different permutations of $n$ distinct objects taken $r$ at a time is
$$\frac{n!}{(n-r)!}$$

Option B is the number of different combinations of $n$ distinct objects taken $r$ at a time. Option C is the number of different permutations of $n$ objects taken $n$ at a time, given that $n$ are of type $i$, where $i = 1, 2, \ldots, k$ and $\sum n_i = n$. Option D is the median of discrete data arranged in increasing order when $n$ is odd.

**Answer is A.**

**10.** The equation $y = ax^3 + bx^2 + cx + d$ is a third degree polynomial, which will exactly fit four constraints on a curve. Constraints can be a point, angle, or curvature (i.e., the reciprocal of the radius).

**Answer is D.**

**11.** The mean is the sum of all measurements divided by the number of measurements.

$$\bar{x} = \sum \frac{x_i}{n} = \frac{9117.0}{100} = 91.17$$

**Answer is D.**

**12.** Protists (algae, fungi, and protozoa) are part of the eucaryote group.

**Answer is B.**

**13.** For exponential growth, the constant specific growth rate is

$$X_1 = 10^3 \text{ mg/L at } t_1 = 0$$
$$X_2 = 10^5 \text{ mg/L at } t_2 = 15 \text{ hr}$$

$$\mu = \left(\frac{1}{X}\right)\left(\frac{dX}{dt}\right)$$

Separate the variables and integrate.

$$\mu \, dt = \frac{dX}{X}$$
$$\mu(t_2 - t_1) = \ln X_2 - \ln X_1$$
$$\mu = \frac{\ln X_2 - \ln X_1}{t_2 - t_1}$$
$$= \frac{\ln\left(10^5 \frac{\text{mg}}{\text{L}}\right) - \ln\left(10^3 \frac{\text{mg}}{\text{L}}\right)}{15 \text{ hr} - 0 \text{ hr}}$$
$$= 0.307 \quad (0.3)$$

**Answer is C.**

**14.**

*step 1:*   Find the chronic daily intake (CDI) for adults and children for each time period.

Given a total lifespan (AT) of 70 yr and an exposure factor (EF) of 1, use the EPA Standard Intake Values table and the following air/water exposure equation to calculate the receptor dose.

$$\text{CDI} = \frac{(E)(\text{CR})(\text{EF})(\text{ED})}{(\text{BW})(\text{AT})}$$

$$= \frac{\left(50 \times 10^{-3} \frac{\text{mg}}{\text{L}}\right)\left(1 \frac{\text{L}}{\text{d}}\right)(1)(20 \text{ yr})}{(10 \text{ kg})(70 \text{ yr})}$$

$$= 1.43 \times 10^{-3} \text{ mg/kg·d}$$

*step 2:*   Find the probable risk of additional cancers, $R$, for adults and children for the 20-yr period.

Given a slope factor for orally ingested benzene of $2.0 \times 10^{-2}$ $(\text{mg/kg·d})^{-1}$, use the lifetime risk probability equation.

$$R = (\text{SF})(\text{CDI})$$

$$= \left(2 \times 10^{-2} \left(\frac{\text{mg}}{\text{kg·d}}\right)^{-1}\right)\left(1.43 \times 10^{-3} \frac{\text{mg}}{\text{kg·d}}\right)$$

$$= 2.86 \times 10^{-5}$$

These calculations yield the following probabilities of excess cancer risk.

**Answer is C.**

**15.** $C_j = \text{electricity} + \dfrac{\text{parts}}{\text{maintenance}} + \text{insurance}$

$$= \$300 + \$100 + (0.02)(\$10{,}000)$$
$$= \$600$$

The annual cost of ownership is the equivalent uniform annual cost.

$$\text{EUAC} = C(A/P, 6\%, 8) - S_8(A/F, 6\%, 8) + C_j$$
$$= (\$10{,}000)(0.1610) - (\$1000)(0.1010) + \$600$$
$$= \$2109 \quad (\$2100)$$

**Answer is B.**

**16.** Let $R_j$ be the minimum acceptable average annual return.

$$\text{present worth} = 0$$
$$0 = -C - C_j(P/A, 6\%, 8) + S_8(P/F, 6\%, 8)$$
$$\quad + R_j(P/A, 6\%, 8)$$
$$0 = -\$10{,}000 - (\$600)(6.2098) + (\$1000)(0.6274)$$
$$\quad + R_j(6.2098)$$
$$R_j = \$2109 \quad (\$2110)$$

Note that this is the same as the effective uniform annual cost of ownership.

**Answer is B.**

**17.** $P = -C - C_j(P/A, 6\%, 8) + S_8(P/F, 6\%, 8)$
$$\quad + R_j(P/A, 6\%, 8)$$
$$= -\$10{,}000 - (\$600)(6.2098) + (\$1000)(0.6274)$$
$$\quad + (\$3000)(6.2098)$$
$$= \$5531 \quad (\$5530)$$

**Answer is A.**

**18.** The equivalent uniform annual cost (EUAC) is the sum of the equivalent annual values of its initial and maintenance costs, minus the equivalent annual value of its salvage value.

$$\text{EUAC} = (\$10,000)(A/P, 4\%, 15) + \$450$$
$$- (\$2500)(A/F, 4\%, 15)$$
$$= (\$10,000)(0.1193) + \$450 - (\$2000)(0.0593)$$
$$= \$1524.40 \quad (\$1500)$$

**Answer is C.**

**19.** The straight line depreciation value is equal to the initial cost minus the salvage value, divided by the lifetime in years.

$$D = \frac{C - S_n}{n}$$

Rearrange to calculate the salvage value.

$$S_n = C - Dn = \$14,000 - (\$850)(15)$$
$$= \$1250$$

**Answer is B.**

**20.** The salary levels represent a geometric sequence. Let $S_i$ be the salary at level $i$.

$$S_3 = 1.04 S_2$$
$$S_4 = 1.04 S_3$$
$$S_5 = 1.04 S_4$$
$$S_6 = 1.04 S_5 = (1.04)^4 S_2$$
$$S_2 = S_6 - \frac{\$140}{\text{mo}}$$

$$S_6 = (1.04)^4 \left( S_6 - \frac{\$140}{\text{mo}} \right) = \frac{-(1.04)^4 \left( \frac{\$140}{\text{mo}} \right)}{1 - (1.04)^4}$$

$$= \$964.22 \quad (\$960)$$

**Answer is B.**

**21.**

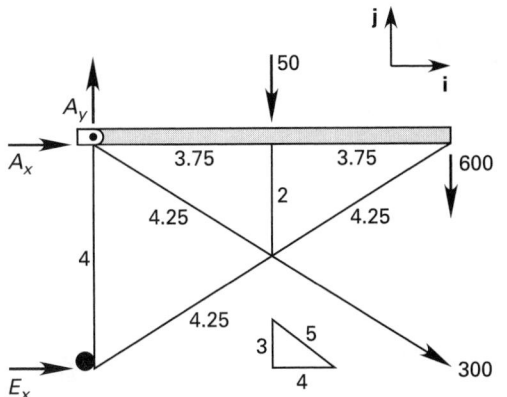

$$\sum F_y = A_y - 50 - 600 - (300) \left( \tfrac{3}{5} \right) = 0$$
$$A_y = 830 \quad \text{[upward]}$$
$$\sum M_A = -(3.75)(50) - (7.5)(600) + (4)(E_x)$$
$$+ (2)(300) \left( \tfrac{4}{5} \right) - (3.75)(300) \left( \tfrac{3}{5} \right) = 0$$
$$E_x = 1220.6 \quad \text{[to the right]}$$
$$\sum F_x = A_x + 1220.6 + (300) \left( \tfrac{4}{5} \right) = 0$$
$$A_x = -1460.6 = 14601 \quad \text{[to the left]}$$

**Answer is B.**

**22.** At point E,

$$\sum F_x = F_{\text{ED}} \left( \frac{7.5}{8.5} \right) + E_x = 0$$

$$F_{\text{ED}} \left( \frac{7.5}{8.5} \right) + 1220.6 = 0$$

$$F_{\text{ED}} = -1383.3 \quad \text{[compression]}$$

$$\sum F_y = F_{\text{ED}} \left( \frac{2}{4.25} \right) + F_{\text{AE}} = 0$$

$$F_{\text{AE}} = -(-1383.3) \left( \frac{2}{4.25} \right)$$

$$= 651 \quad \text{[tension]}$$

At point A,

$$\sum F_y = -F_{\text{AE}} - F_{\text{AD}} \left( \frac{2}{4.25} \right) + A_y = 0$$

$$F_{\text{AD}} = (830 - 651) \left( \frac{4.25}{2} \right) = 380.4 \quad \text{[tension]}$$

(Member ABC is horizontal and has no vertical force component.)

**Answer is C.**

**23.** A horizontal force to the right will counter the other moments.

Assume $E_x = 0$.

$$\sum M_A = -(3.75)(50) - (7.5)(600) + (2)(300) \left( \tfrac{4}{5} \right)$$
$$- (3.75)(300) \left( \tfrac{3}{5} \right) + 2F_{\text{D}}' = 0$$
$$F_{\text{D}}' = 2441.3 \quad (2441 \quad \text{[to the right]})$$

**Answer is D.**

**24.** Draw free body diagrams for joints F, G, and D.

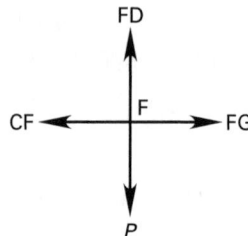

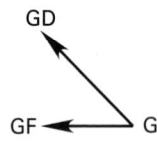

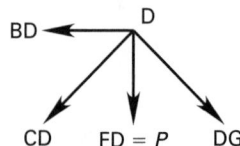

Sum the forces in the $y$-direction for the F joint.

$$FD - P = 0$$
$$FD = P$$

Sum the forces in the $y$-direction for the G joint, (GD = DG).

$$DG\left(\frac{3}{5}\right) = 0$$
$$DG = 0$$

Sum the forces in the $y$-direction for the D joint.

$$P = \left(\frac{-3}{5}\right)DG - \left(\frac{3}{5}\right)CD$$
$$= \left(\frac{-3}{5}\right)(0) - \left(\frac{3}{5}\right)CD$$
$$= \frac{-3}{5}CD$$
$$CD = \frac{-5}{3}P \qquad \text{[Eq. 1]}$$

Sum the forces in the $x$-direction.

$$BD + \left(\frac{4}{5}\right)CD - \left(\frac{4}{5}\right)DG = 0$$
$$BD + \left(\frac{4}{5}\right)CD - \left(\frac{4}{5}\right)(0) = 0$$
$$BD = \frac{-4}{5}\left(\frac{-5}{3}P\right) \qquad \text{[Eq. 2]}$$

Combining Eqs. 1 and 2,

$$BD = \frac{-4}{5}\left(\frac{-5}{3}P\right)$$
$$= \left(\frac{4}{3}\right)P$$

The maximum load member BD can bear without buckling is

$$P_{cr} = \frac{\pi^2 EI}{(KL)^2}$$

$$K = 1 \text{ for pinned ends}$$

$$\left(\frac{4}{3}\right)P = \frac{\pi^2(42.0\text{ GPa})\left(\frac{1\times10^9\text{ Pa}}{\text{GPa}}\right)\left(\frac{1}{4}\right)\pi}{\times\left(1.25\text{ cm}\times\frac{\text{m}}{100\text{ cm}}\right)^4}{((1)(4\text{ m}))^2}$$

$$P = 496.8\text{ N}\quad(500\text{ N})$$

**Answer is C.**

**25.** Sum the moments at A.

$$M_0 - R_BL = 0$$
$$R_B = \frac{M_0}{L} = \frac{100\text{ N·m}}{10\text{ m}}$$
$$= 10\text{ N}$$
$$R_A = \frac{M_0}{L} = 10\text{ N}$$

The shear and moment diagrams are shown.

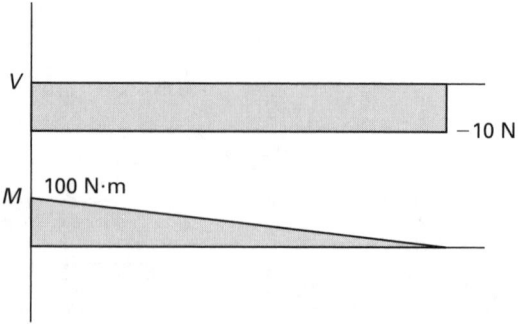

$$EI\frac{d^3y}{dx^3} = V(x) = \frac{-100\text{ N·m}}{10\text{ m}}$$
$$= -10\text{ N}$$
$$EI\frac{d^2y}{dx^2} = M(x) = (-10\text{ N})(x) + C_1$$

However, at $x = 0$, $M = M_0$, and $C_1 = M_0$,

$$EI\frac{dy}{dx} = EI\theta(x)$$
$$= \frac{(-10 \text{ N})(x^2)}{2} + (100 \text{ N·m})x + C_2$$

Nothing is known about the slope at any point.

$$EIy(x) = (-10 \text{ N})\left(\frac{x^3}{6}\right) + (100 \text{ N·m})\left(\frac{x^2}{2}\right)$$
$$+ C_2x + C_3$$

At $x = 0$, $y = 0$, so $C_3 = 0$.

Also, at $x = 10$ m, $y = 0$ m,

$$0 = \frac{(-10 \text{ N})(10 \text{ m})^3}{6} + \frac{(100 \text{ N·m})(10 \text{ m})^2}{2}$$
$$+ C_2(10 \text{ m})$$
$$C_2 = \frac{-1000}{3} \text{ N·m}^2$$

At the midspan, $x = 5$.

$$EIy(5) = (-10 \text{ N})\frac{(5 \text{ m})^3}{6} + \frac{(100 \text{ N·m})(5 \text{ m})^2}{2}$$
$$- \frac{1000 \text{ N·m}^2}{3}(5 \text{ m}) + 0$$
$$y = \frac{-625}{EI} \text{ m}$$

**Answer is A.**

**26.** The relationship between the modulus of elasticity, shear modulus, and Poisson's ratio is commonly expressed in one of two ways.

$$G = \frac{E}{2(1 + \nu)}$$
$$E = 2G(1 + \nu)$$

**Answer is A.**

**27.** Calculate the area of the cross section of the pipe.

$$A = \pi\left(\left(\frac{d_{\text{out}}}{2}\right)^2 - \left(\frac{d_{\text{in}}}{2}\right)^2\right)$$
$$= \pi\left(\left(\frac{5 \text{ cm}}{2}\right)^2 - \left(\frac{1 \text{ cm}}{2}\right)^2\right)$$
$$= 18.85 \text{ cm}^2$$
$$\sigma = \frac{P}{A}$$
$$= \left(\frac{14.4 \text{ N}}{18.85 \text{ cm}^2}\right)\left(100 \frac{\text{cm}}{\text{m}}\right)^2$$
$$= 76.39 \text{ Pa} \quad (76 \text{ Pa})$$

**Answer is C.**

**28.** Hooke's law for shear is

$$\gamma = \frac{\tau_{xy}}{G} = \frac{12\,000 \text{ kPa}}{87 \times 10^6 \text{ kPa}}$$
$$= 1.38 \times 10^{-4} \text{ rad} \quad (1.4 \times 10^{-4} \text{ rad})$$

**Answer is B.**

**29. Answer is A.**

**30.** Cold working, or plastically deforming, a metal increases the strength and decreases the ductility.

**Answer is B.**

**31.** This illustration depicts a cup-and-cone failure, typical of moderately ductile materials. Ductility is the ratio of ultimate to yield strain.

**Answer is B.**

**32.** The strain at failure is found by extending a line from the failure point to the strain axis parallel to the linear portion of the curve (i.e., the portion of the line between the origin and the proportionality limit).

$$\text{percent elongation} = \varepsilon_f \times 100\%$$
$$= 0.25 \times 100\%$$
$$= 25\%$$

Note that the percent elongation is an indicator of the ductility of a metal, but it is not the same as the ductility.

**Answer is C.**

**33.** Creep is defined as the continuous yielding of material under constant stress.

**Answer is C.**

**34.** By definition, the mass of an atom is its atomic weight divided by Avogadro's number. From the periodic table, the atomic weight of nickel is 58.69.

$$m = \frac{58.69 \frac{\text{g}}{\text{mol}}}{6.02 \times 10^{23} \frac{\text{atoms}}{\text{mol}}}$$
$$= 9.749 \times 10^{-23} \text{ g/atom} \quad (9.7 \times 10^{-23} \text{ g/atom})$$

**Answer is B.**

**35.** The mass of the bar is

$$m = \rho V = \rho A L$$

$$= \left(2500 \ \frac{\text{kg}}{\text{m}^3}\right)(150 \ \text{cm}^2)\left(\frac{1 \ \text{m}}{100 \ \text{cm}}\right)^2(2 \ \text{m})$$

$$= 75 \ \text{kg}$$

The total gravitational force is experienced by the metal at the suspension point. Farther down the rod there is less volume contributing to the force, and the stress is reduced. The average force on the metal in the bar is half of the maximum value.

$$F_{\text{ave}} = \left(\frac{1}{2}\right) F_{\text{max}} = \left(\frac{1}{2}\right) mg$$

$$= \left(\frac{1}{2}\right)(75 \ \text{kg})\left(9.81 \ \frac{\text{m}}{\text{s}^2}\right)$$

$$= 367.87 \ \text{N}$$

The elongation is

$$\Delta L = \varepsilon L_o = \left(\frac{\sigma}{E}\right) L_o$$

$$= \left(\frac{F}{AE}\right) L_o$$

$$= \frac{(367.87 \ \text{N})(2 \ \text{m})}{\left(\begin{array}{c}(150 \ \text{cm}^2)\left(\dfrac{1 \ \text{m}}{100 \ \text{cm}}\right)^2(210 \ \text{GPa}) \\ \times \left(1.0 \times 10^9 \ \dfrac{\text{Pa}}{\text{GPa}}\right)\end{array}\right)}$$

$$= 2.336 \times 10^{-7} \ \text{m} \quad (2.34 \times 10^{-7} \ \text{m})$$

**Answer is D.**

**36.** $F_x = \overline{p}A = \gamma \overline{h} A = \rho g \overline{h} A$

$$= \left(1000 \ \frac{\text{kg}}{\text{m}^3}\right)\left(9.81 \ \frac{\text{m}}{\text{s}^2}\right)\left(\tfrac{1}{2}\right)(6 \ \text{m} + 8 \ \text{m})(2 \ \text{m})$$

$$= 137\,340 \ \text{N/m}$$

$F_y = $ weight of water above gate $= \rho g V$

$$= \left(1000 \ \frac{\text{kg}}{\text{m}^3}\right)\left(9.81 \ \frac{\text{m}}{\text{s}^2}\right)$$

$$\times \left((2 \ \text{m})(6 \ \text{m}) + \left(\tfrac{1}{2}\right)(2 \ \text{m})(2 \ \text{m})\right)$$

$$= 137\,340 \ \text{N/m}$$

$$F = \sqrt{F_x^2 + F_y^2}$$

$$= \sqrt{\left(137\,340 \ \frac{\text{N}}{\text{m}}\right)^2 + \left(137\,340 \ \frac{\text{N}}{\text{m}}\right)^2}$$

$$= 194\,000 \ \text{N/m} \quad (190 \ \text{kN/m})$$

**Answer is B.**

**37.** $D = (17 \ \text{cm})\left(\dfrac{1 \ \text{m}}{100 \ \text{cm}}\right)$

$$= 0.17 \ \text{m}$$

$$A = \frac{\pi D^2}{4} = \frac{\pi (0.17 \ \text{m})^2}{4}$$

$$= 0.0227 \ \text{m}^2$$

$$v = \frac{Q}{A}$$

$$= \frac{\left(3300 \ \dfrac{\text{L}}{\text{min}}\right)\left(\dfrac{1 \ \text{min}}{60 \ \text{s}}\right)\left(0.001 \ \dfrac{\text{m}^3}{\text{L}}\right)}{0.0227 \ \text{m}^2}$$

$$= 2.42 \ \text{m/s}$$

$$h_f = \frac{fLv^2}{2Dg}$$

$$= \frac{(0.03)(1 \ \text{km})\left(1000 \ \dfrac{\text{m}}{\text{km}}\right)\left(2.42 \ \dfrac{\text{m}}{\text{s}}\right)^2}{(2)(0.17 \ \text{m})\left(9.81 \ \dfrac{\text{m}}{\text{s}^2}\right)}$$

$$= 52.7 \ \text{m} \quad (53 \ \text{m})$$

**Answer is C.**

**38.** $A_1 = \dfrac{\pi d_1^2}{4} = \dfrac{\pi\left((5 \ \text{cm})\left(\dfrac{1 \ \text{m}}{100 \ \text{cm}}\right)\right)^2}{4}$

$$= 0.001963 \ \text{m}^2$$

$$A_2 = \frac{\pi d_2^2}{4} = \frac{\pi\left((1.8 \ \text{cm})\left(\dfrac{1 \ \text{m}}{100 \ \text{cm}}\right)\right)^2}{4}$$

$$= 2.545 \times 10^{-4} \ \text{m}^2$$

$$v_1 = \frac{\dot{m}}{\rho A_1} = \frac{8.0 \ \dfrac{\text{kg}}{\text{s}}}{\left(1000 \ \dfrac{\text{kg}}{\text{m}^3}\right)(0.001963 \ \text{m}^2)}$$

$$= 4.07 \ \text{m/s}$$

$$v_2 = \frac{\dot{m}}{\rho A_2} = \frac{8.0 \ \dfrac{\text{kg}}{\text{s}}}{\left(1000 \ \dfrac{\text{kg}}{\text{m}^3}\right)(2.545 \times 10^{-4} \ \text{m}^2)}$$

$$= 31.43 \ \text{m/s}$$

$$p_1 - p_2 = \left(\frac{\rho}{2}\right)(v_2^2 - v_1^2)$$

$$= \left(\frac{1000 \ \dfrac{\text{kg}}{\text{m}^3}}{2}\right)\left(\left(31.43 \ \frac{\text{m}}{\text{s}}\right)^2 - \left(4.075 \ \frac{\text{m}}{\text{s}}\right)^2\right)$$

$$\times \left(\frac{1 \ \text{kPa}}{1000 \ \text{Pa}}\right)$$

$$= 486 \ \text{kPa} \quad (490 \ \text{kPa})$$

**Answer is B.**

**39.** The static pressure is

$$p_s = (60.96 \text{ cm}) \left( 1000 \ \frac{\text{kg}}{\text{m}^3} \right) \left( 9.81 \ \frac{\text{m}}{\text{s}^2} \right) \left( \frac{1 \text{ m}}{100 \text{ cm}} \right)$$

$$+ \, 10\,342 \text{ Pa}$$

$$= 16\,322 \text{ Pa}$$

The stagnation pressure is

$$p_0 = \left( 0.1336 \ \frac{\text{N}}{\text{cm}^3} \right) (25.4 \text{ cm}) \left( 100 \ \frac{\text{cm}}{\text{m}} \right)^2$$

$$- \left( 9810 \ \frac{\text{N}}{\text{m}^3} \right) \left( \frac{1 \text{ m}}{100 \text{ cm}} \right) (50.8 \text{ cm} + 25.4 \text{ cm})$$

$$= 26\,459 \text{ Pa}$$

$$\text{v} = \sqrt{\frac{(2)(p_0 - p_s)}{\rho}}$$

$$= \sqrt{\frac{(2)(26\,459 \text{ Pa} - 16\,322 \text{ Pa})}{1000 \ \dfrac{\text{kg}}{\text{m}^3}}}$$

$$= 4.5 \text{ m/s}$$

**Answer is C.**

**40.** The potential energy is zero at the centerline of the pump. The pressure energy is

$$E_p = \frac{p}{\rho} = \frac{(1000 \text{ kPa}) \left( 1000 \ \dfrac{\text{Pa}}{\text{kPa}} \right)}{1000 \ \dfrac{\text{kg}}{\text{m}^3}}$$

$$= 1000 \text{ J/kg}$$

The velocity energy is

$$E_v = \frac{\text{v}^2}{2} = \frac{\left( 2 \ \dfrac{\text{m}}{\text{s}} \right)^2}{2}$$

$$= 2 \text{ J/kg}$$

The total energy is the sum of the pressure energy and the velocity energy, as follows.

$$E_t = E_p + E_v = 1000 \ \frac{\text{J}}{\text{kg}} + 2 \ \frac{\text{J}}{\text{kg}}$$

$$= 1002 \text{ J/kg}$$

The available energy at the surface is only potential energy.

$$E_{t_2} = E_{t_1} = E_{p_2}$$

$$z_2 g = 1002 \text{ J/kg}$$

So,

$$z_2 = \frac{E_{t_2}}{g} = \frac{1002 \ \dfrac{\text{J}}{\text{kg}}}{9.81 \ \dfrac{\text{m}}{\text{s}^2}}$$

$$= 102.1 \text{ m}$$

**Answer is A.**

**41.** The hydraulic radius is

$$R = \frac{D}{4} = \frac{0.5 \text{ m}}{4}$$

$$= 0.125 \text{ m}$$

The flow rate is

$$Q = A\text{v} \left( \frac{1}{n} \right) A R^{2/3} \sqrt{S}$$

$$= \left( \frac{1}{0.015} \right) \left( \frac{\pi (0.5 \text{ m})^2}{4} \right) (0.125 \text{ m})^{2/3} \sqrt{0.001}$$

$$= 0.103 \text{ m}^3/\text{s} \quad (0.1 \text{ m}^3/\text{s})$$

**Answer is A.**

**42.** Because the channel bottom is level on both sides of the sluice gate, $z_1 = z_2$. Reducing Bernoulli's equation gives

$$d_1 + \frac{\text{v}_1^2}{2g} = d_2 + \frac{\text{v}_2^2}{2g}$$

or

$$2 \text{ m} + \frac{\text{v}_1^2}{2g} = 0.6 \text{ m} + \frac{\text{v}_2^2}{2g}$$

$\text{v}_1$ and $\text{v}_2$ are related by continuity.

$$Q_1 = Q_2$$

$$A_1 \text{v}_1 = A_2 \text{v}_2$$

Solving gives

$$(2 \text{ m})(4 \text{ m})\text{v}_1 = (0.6 \text{ m})(4 \text{ m})\text{v}_2$$

$$\text{v}_1 = 0.3\text{v}_2$$

Substitute this into the Bernoulli equation.

$$2 \text{ m} + \frac{(0.3\text{v}_2)^2}{(2) \left( 9.81 \ \dfrac{\text{m}}{\text{s}^2} \right)} = 0.6 \text{ m} + \frac{\text{v}_2^2}{(2) \left( 9.81 \ \dfrac{\text{m}}{\text{s}^2} \right)}$$

$$\text{v}_2 = 5.49 \text{ m/s} \quad (5.5 \text{ m/s})$$

**Answer is D.**

**43.** First, find the area of the channel.

$$A = dw = (3 \text{ m})(5 \text{ m})$$

$$= 15 \text{ m}^2$$

Find the hydraulic radius.

$$R = \frac{dw}{w + 2d}$$

$$= \frac{(3 \text{ m})(5 \text{ m})}{5 \text{ m} + (2)(3 \text{ m})}$$

$$= 1.36 \text{ m}$$

Use Manning's equation.

$$Q = \left(\frac{1}{n}\right) AR^{2/3}\sqrt{S}$$

$$= \left(\frac{1}{0.013}\right)(15 \text{ m}^2)(1.36 \text{ m})^{2/3}\sqrt{0.004}$$

$$= 89.59 \text{ m}^3/\text{s}$$

Use the continuity equation and solve for v.

$$Q = vA$$

$$v = \frac{Q}{A} = \frac{89.59 \text{ m}^3/\text{s}}{15 \text{ m}^2}$$

$$= 5.97 \text{ m/s} \quad (6 \text{ m/s})$$

**Answer is B.**

**44.**  Begin by finding the hydraulic radius.

$$R = \frac{A}{P}$$

$$= \frac{(2.4 \text{ m})(1.5 \text{ m})}{1.5 \text{ m} + 2.4 \text{ m} + 1.5 \text{ m}}$$

$$= 0.667 \text{ m}$$

Use the Manning's equation to solve for flow.

$$Q = \frac{1}{n}AR^{2/3}\sqrt{S}$$

$$= \left(\frac{1}{0.012}\right)((2.4 \text{ m})(1.5 \text{ m}))(0.667 \text{ m})^{2/3}\sqrt{0.002}$$

$$= 10.2 \text{ m}^3/\text{s} \quad (10 \text{ m}^3/\text{s})$$

**Answer is C.**

**45.**        $0.8 = \text{power factor} = \cos\theta$

$$\theta = \arccos 0.8 = 36.87°$$

The leading power factor is equal to the positive angle for the current. The impedance angle is the negative of this. The load impedance in phasor form is

$$\mathbf{Z}_l = 55\angle-36.87° \ \Omega$$

The secondary voltage is 2000 V, and for reference to the secondary side, the phase offset is 0°.

$$\mathbf{E}_2 = 2000\angle0° \text{ V}$$

$$\mathbf{I}_2 = \frac{\mathbf{E}_2}{\mathbf{Z}_2} = \frac{2000\angle0° \text{ V}}{55\angle-36.87° \ \Omega}$$

$$= 36.36\angle36.87° \text{ A}$$

**Answer is D.**

**46.**        $\mathbf{E}_2 = 2000\angle0° \text{ V} \quad (2.0\angle0° \text{ kV})$

**Answer is A.**

**47.**  Referred to the primary side, the equivalent circuit diagram is

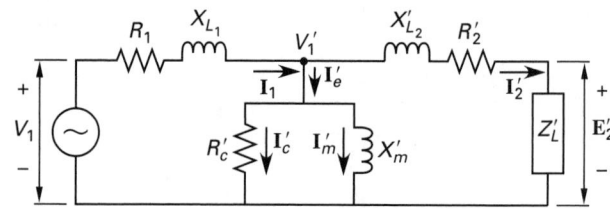

$$\mathbf{I}_2' = \left(\frac{N_2}{N_1}\right)\mathbf{I}_2 = \left(\frac{2000 \text{ turns}}{10,000 \text{ turns}}\right)(36.36\angle36.87° \text{ A})$$

$$= 7.27\angle36.87° \text{ A}$$

**Answer is C.**

**48.**        $\mathbf{E}_2' = \left(\frac{N_1}{N_2}\right)\mathbf{E}_2$

$$= \left(\frac{10,000 \text{ turns}}{2000 \text{ turns}}\right)(2000\angle0° \text{ V})$$

$$= 10,000\angle0° \text{ V} \quad (10.0\angle0° \text{ kV})$$

**Answer is C.**

**49.**        $\dfrac{R(s)}{F(s)} = T(s) = \dfrac{7}{s^2 + 4s + 3}$

$$F(t) = \delta(t)$$

$$F(s) = 1$$

$$R(s) = T(s)F(s) = \frac{7}{s^2 + 4s + 3}$$

$$= \frac{-\dfrac{7}{2}}{s + 3} + \dfrac{\dfrac{7}{2}}{s + 1}$$

$$R(t) = \left(\frac{7}{2}\right)\left(e^{-t} - e^{-3t}\right)$$

**Answer is D.**

**50.** $F(t) = 5u(t)$

$$F(s) = \frac{5}{s}$$

$$R(s) = \frac{35}{(s^2 + 4s + 3)s}$$

$$R(t)\Big|_{t\to\infty} = sR(s)\Big|_{s\to 0} = \frac{35}{0 + (4)(0) + 3} = 35/3$$

**Answer is C.**

**51.** The input phasor is $\mathbf{F} = 7\angle 60°$ with $\omega = 2$.

$$\mathbf{T}(j\omega) = \mathbf{T}(j2) = \frac{7}{(j2)^2 + (4)(j2) + 3} = \frac{7}{-1 + 8j}$$

$$= \frac{7}{8.062\angle 97.13°}$$

$$= 0.868\angle -97.13°$$

$$\mathbf{R}(j\omega) = \mathbf{F}(j\omega)\mathbf{T}(j\omega) = (7\angle 60°)(0.868\angle -97.13°)$$

$$= 6.08\angle -37.13° \quad (6.01\angle -37°) \quad [\text{at } \omega = 2]$$

**Answer is B.**

**52.** $\text{COP}_{\text{refrigeration}}$

$$= \frac{T_L}{T_H - T_L}$$

$$= \frac{-123.3°C + 273}{(-23.3°C + 273) - (-123.3°C + 273)}$$

$$= 1.5$$

**Answer is D.**

**53.** $W = mRT \ln\frac{1}{2}$

$$= \frac{(1 \text{ kg})\left(278 \frac{J}{\text{kg·K}}\right)(273\text{K})\left(\ln\frac{1}{2}\right)}{1000 \frac{J}{\text{kJ}}}$$

$$= -52.6 \text{ kJ} \quad (53 \text{ kJ})$$

**Answer is B.**

**54.** At 1 MPa and 500°C,

$$h_3 = 3478.5 \text{ kJ/kg}$$

At 9.59 kPa,

$$h_f = 188.45 \text{ kJ/kg}$$
$$h_{fg} = 2394.8 \text{ kJ/kg}$$
$$h_4 = 188.45 \frac{\text{kJ}}{\text{kg}} + (0.98)\left(2394.8 \frac{\text{kJ}}{\text{kg}}\right)$$
$$= 2535.4 \text{ kJ/kg}$$

$$W_{\text{actual}} = h_3 - h_4 = 3478.5 \frac{\text{kJ}}{\text{kg}} - 2535.4 \frac{\text{kJ}}{\text{kg}}$$
$$= 943.1 \text{ kJ/kg}$$

$$W_{\text{ideal}} = \frac{W_{\text{actual}}}{\eta} = \frac{943.1 \frac{\text{kJ}}{\text{kg}}}{0.857}$$
$$= 1100 \text{ kJ/kg}$$

**Answer is D.**

**55.** $$\dot{m} = \frac{P}{W_{\text{actual}}} = \frac{(1 \times 10^6 \text{ W})\left(3600 \frac{\text{s}}{\text{h}}\right)}{\left(943.1 \frac{\text{kJ}}{\text{kg}}\right)\left(1000 \frac{\text{J}}{\text{kJ}}\right)}$$

$$= 3817 \text{ kg/h} \quad (3800 \text{ kg/h})$$

**Answer is C.**

**56.** The water will leave the condenser approximately saturated at 9.59 kPa. From the steam table,

$$h_5 \approx 188.45 \text{ kJ/kg} \quad (190 \text{ kJ/kg})$$

**Answer is C.**

**57.** $$h_4 = 2535.4 \text{ kJ/kg}$$

From an energy balance around the condenser,

$$\Delta h_{\text{cooling water}} = \Delta h_{\text{steam}}$$
$$\dot{m}_{\text{cooling water}}c_p\Delta T = \dot{m}_{\text{steam}}(h_4' - h_5)$$
$$\dot{m}_{\text{cooling water}}\left(4180 \frac{\text{J}}{\text{kg·°C}}\right)(43°C - 20°C)$$
$$= \left(4454 \frac{\text{kg}}{\text{h}}\right)\left(2535.4 \frac{\text{kJ}}{\text{kg}} - 188.45 \frac{\text{kJ}}{\text{kg}}\right)\left(1000 \frac{\text{J}}{\text{kJ}}\right)$$
$$\dot{m}_{\text{cooling water}} = 1.09 \times 10^5 \text{ kg/h} \quad (1.1 \times 10^5 \text{ kg/h})$$

**Answer is C.**

**58.** For saturated water at 9.593 kPa,

$$v_5 = 0.00101 \text{ m}^3/\text{kg}$$
$$P_1 = P_3 = 1 \text{ MPa}$$

For an incompressible fluid, the pump work is

$$W_{\text{ideal}} = v_5(P_1 - P_5)$$
$$= \left(0.00101 \frac{\text{m}^3}{\text{kg}}\right)$$
$$\times (1 \times 10^6 \text{ Pa} - 9.593 \times 10^3 \text{ Pa})$$
$$= 1000 \frac{\text{J}}{\text{kg}}$$

$$W_{\text{actual}} = \frac{W_{\text{ideal}}}{\eta_{\text{pump}}} = \frac{1000 \frac{\text{J}}{\text{kg}}}{0.60}$$
$$= 1667 \text{ J/kg} \quad (1700 \text{ J/kg})$$

**Answer is C.**

**59.** The saturation temperature corresponding to 9.593 kPa is 45°C.

$$T_1 = T_5 + \frac{W_{\text{pump}}}{c_p}$$

$$= 45°C + \frac{1667 \text{ W}}{4180 \frac{\text{J}}{\text{kg·°C}}}$$

$$= 45.4°C$$

**Answer is B.**

**60.** For an isentropic process,

$$T_2 = T_1 \left(\frac{p_2}{p_1}\right)^{(k-1)/k}$$

$$= (355\text{K}) \left(\frac{10.73 \frac{\text{N}}{\text{cm}^2}}{7.25 \frac{\text{N}}{\text{cm}^2}}\right)^{(1.4-1)/1.4}$$

$$= 397.08\text{K} \quad (400\text{K})$$

**Answer is C.**

# Exam 2—Answer Key

## Morning Section

| | | | | | |
|---|---|---|---|---|---|
| 1. B | 21. D | 41. C | 61. C | 81. B | 101. B |
| 2. B | 22. B | 42. B | 62. D | 82. A | 102. C |
| 3. A | 23. D | 43. D | 63. A | 83. B | 103. D |
| 4. D | 24. B | 44. A | 64. B | 84. C | 104. C |
| 5. A | 25. A | 45. B | 65. D | 85. A | 105. C |
| 6. D | 26. B | 46. A | 66. B | 86. B | 106. B |
| 7. C | 27. B | 47. C | 67. C | 87. D | 107. C |
| 8. C | 28. D | 48. C | 68. A | 88. D | 108. C |
| 9. B | 29. B | 49. A | 69. C | 89. C | 109. D |
| 10. A | 30. B | 50. B | 70. D | 90. C | 110. B |
| 11. D | 31. B | 51. D | 71. C | 91. B | 111. D |
| 12. B | 32. C | 52. D | 72. D | 92. B | 112. C |
| 13. C | 33. B | 53. B | 73. D | 93. B | 113. C |
| 14. A | 34. D | 54. A | 74. D | 94. B | 114. B |
| 15. A | 35. A | 55. C | 75. D | 95. D | 115. B |
| 16. D | 36. B | 56. A | 76. B | 96. D | 116. A |
| 17. C | 37. C | 57. C | 77. A | 97. B | 117. D |
| 18. B | 38. D | 58. B | 78. A | 98. C | 118. B |
| 19. D | 39. D | 59. C | 79. C | 99. D | 119. D |
| 20. C | 40. B | 60. B | 80. D | 100. C | 120. C |

## Afternoon Section

| | | | | | |
|---|---|---|---|---|---|
| 1. C | 11. D | 21. D | 31. B | 41. C | 51. A |
| 2. B | 12. C | 22. C | 32. D | 42. A | 52. D |
| 3. C | 13. D | 23. C | 33. B | 43. D | 53. A |
| 4. D | 14. D | 24. A | 34. A | 44. B | 54. A |
| 5. C | 15. C | 25. D | 35. D | 45. D | 55. C |
| 6. B | 16. B | 26. A | 36. B | 46. C | 56. D |
| 7. D | 17. B | 27. D | 37. D | 47. A | 57. D |
| 8. B | 18. C | 28. C | 38. A | 48. B | 58. B |
| 9. A | 19. D | 29. D | 39. B | 49. C | 59. D |
| 10. C | 20. C | 30. B | 40. A | 50. A | 60. D |

# Solutions for Exam 2–Morning Section

**1.**
$$m = \frac{dy}{dx}\bigg|_{x=5} = 24x$$
$$= (24)(5)$$
$$= 120$$

**Answer is B.**

**2.** $y = 6 - 2x$ is the equation of the line.
$y \le 6 - 2x$ describes the shaded area.

**Answer is B.**

**3.** Use the substitution method.
$$x = \frac{7}{3} + 2y$$
$$(2)\left(\frac{7}{3} + 2y\right) - 11y = -5$$
$$y = \frac{29}{21}$$
$$x = \frac{7}{3} + (2)\left(\frac{29}{21}\right) = \frac{107}{21}$$

**Answer is A.**

**4.**
$$(\cot^2\theta)(\sin^2\theta) + \frac{1}{\csc^2\theta} = \left(\frac{\cos^2\theta}{\sin^2\theta}\right)(\sin^2\theta) + \sin^2\theta$$
$$= \cos^2\theta + \sin^2\theta$$
$$= 1$$

**Answer is D.**

**5.**
$$\frac{d\sqrt{2x + 9x^2}}{dx} = \left(\tfrac{1}{2}\right)\left(\frac{1}{\sqrt{2x + 9x^2}}\right)(2 + 18x)$$
$$= \frac{1 + 9x}{\sqrt{2x + 9x^2}}$$

**Answer is A.**

**6.** Use L'Hôpital's rule.
$$\frac{\dfrac{d}{dx}(10x^2 - 5x + 1)}{\dfrac{d}{dx}(10x^2 - 6x)} = \frac{20x - 5}{20x - 6}$$

Use L'Hôpital's rule again.
$$\frac{\dfrac{d}{dx}(20x - 5)}{\dfrac{d}{dx}(20x - 6)} = \frac{20}{20} = 1$$

**Answer is D.**

**7.**
$$\int x(x + 1)dx = \int (x^2 + x)dx$$
$$= \frac{x^3}{3} + \frac{x^2}{2} + C$$

**Answer is C.**

**8.** Leading zeros are not significant, so the number has 10 significant digits.

**Answer is C.**

**9.**

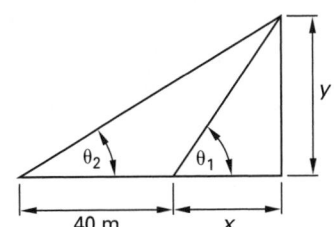

$$\tan\theta_2 = \frac{y}{x + 40 \text{ m}} = \frac{x\tan\theta_1}{x + 40 \text{ m}}$$
$$(x + 40 \text{ m})(\tan\theta_2) = x\tan\theta_1$$
$$\theta_1 = 46.8°$$
$$\theta_2 = 29.23°$$
$$x = \frac{(-40 \text{ m})(\tan\theta_2)}{\tan\theta_2 - \tan\theta_1}$$
$$= 46.4 \text{ m} \quad (46 \text{ m})$$

**Answer is B.**

**10.** $\cos\theta = \dfrac{\mathbf{V}_1 \cdot \mathbf{V}_2}{|\mathbf{V}_1||\mathbf{V}_2|}$
$$= \frac{(3)(2) + (2)(3) + (1)(2)}{\sqrt{(3)^2 + (2)^2 + (1)^2}\sqrt{(2)^2 + (3)^2 + (2)^2}}$$
$$= \frac{14}{\sqrt{14}\sqrt{17}}$$
$$= 0.907485$$
$$\theta = \cos^{-1}(0.907485) = 24.8° \quad (25°)$$

**Answer is A.**

**11.**

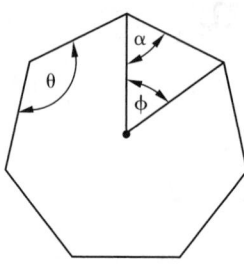

$$\phi = \frac{360°}{7}$$
$$= 51.43°$$

$$\alpha = \frac{180° - \phi}{2}$$
$$= 64.28°$$

$$\theta = 2\alpha$$
$$= 128.6° \quad (129°)$$

**Answer is D.**

**12.** $\qquad \cos x = \sum_{n=0}^{\infty} \left( \frac{x^{2n}}{(2n)!} \right) (-1)^n$

**Answer is B.**

**13.** $\qquad \overline{A} \cup B = (5, 6, 8, 10) \ \cup \ (4, 5, 9, 10)$
$$= (4, 5, 6, 8, 9, 10)$$

**Answer is C.**

**14.** Graph the function on the polar coordinate system.

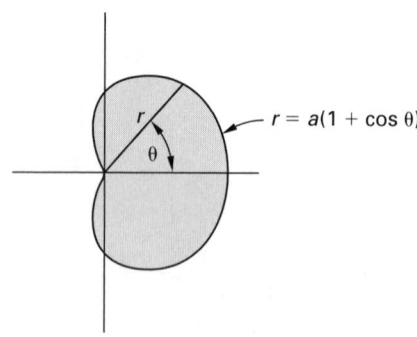

By definition,

$$A = \tfrac{1}{2} \int_0^{2\pi} [f(\theta)]^2 d\theta$$

Since the curve is symmetrical,

$$A = \left(\tfrac{1}{2}\right) (2) \int_0^{\pi} [f(\theta)]^2 d\theta = a^2 \int_0^{\pi} (1 + \cos\theta)^2 d\theta$$
$$= a^2 \int_0^{\pi} (1 + 2\cos\theta + \cos^2\theta) d\theta$$
$$= a^2 \left[ \theta + 2\sin\theta + \tfrac{1}{2}\theta + \tfrac{1}{4}\sin 2\theta \right]_0^{\pi}$$
$$= a^2 \left[ \tfrac{3}{2}\theta + 2\sin\theta + \tfrac{1}{4}\sin 2\theta \right]_0^{\pi}$$
$$= a^2 \left( \frac{3\pi}{2} + 0 + 0 - 0 - 0 - 0 \right) = \frac{3\pi a^2}{2}$$

**Answer is A.**

**15.** Using matrix multiplication,

$$9B_1 + 7B_2 = 2$$
$$B_1 + 3B_2 = 1$$

Solving simultaneously,

$$-20B_2 = -7$$
$$B_2 = 7/20$$
$$B_1 = 1 - 3B_2 = 1 - (3)\left(\frac{7}{20}\right) = -1/20$$

**Answer is A.**

**16.** Divide through by $x$.

$$y' + 3 - \frac{1}{x} = 0$$

Simplify.

$$\frac{dy}{dx} = \frac{1}{x} - 3$$
$$dy = \left( \frac{1}{x} - 3 \right) dx$$

**Answer is D.**

**17.** $\qquad x\frac{dy}{dx} + 3x - 1 = 0$

$$dy = \left( \frac{1 - 3x}{x} \right) dx = \left( \frac{1}{x} - 3 \right) dx$$
$$y = \ln x - 3x + C$$

**Answer is C.**

**18.** This is the vector triple scalar product, which is zero only if $\overline{A}$ lies in the same plane as $\overline{B}$ and $\overline{C}$.

**Answer is B.**

**19.**
$$\overline{x} = \frac{\sum x_i}{n} = \frac{30.46}{7}$$
$$= 4.351$$

$$s = \sqrt{\frac{\sum (x_i - \overline{x})^2}{n-1}} = \sqrt{\frac{0.005887}{7-1}}$$
$$= 0.03132 \quad (0.0313)$$

**Answer is D.**

**20.** The number of combinations of $r$ items selected from a set of $n$ items is

$$C(n,r) = \frac{n!}{(n-r)!r!} \quad \text{[for } r \leq n\text{]}$$

$$C(6,4) = \frac{6!}{(6-4)!4!} = 15$$

**Answer is C.**

**21.** First calculate the probability of no heads, and then subtract that from 1 to get the probability of at least one head.

$$p\left\{\begin{array}{c}\text{three tosses,} \\ \text{three tails}\end{array}\right\} = p\{\text{one toss, tails}\}^3$$
$$= \left(\frac{1}{2}\right)^3$$
$$= 0.125$$

$$p\left\{\begin{array}{c}\text{three tosses, at} \\ \text{least one head}\end{array}\right\} = 1 - p\{\text{three tosses, three tails}\}$$
$$= 1 - 0.125$$
$$= 0.875 \quad (0.88)$$

**Answer is D.**

**22.** The number of ring permutations of $n$ items is

$$P_{\text{ring}}(n,n) = \frac{P(n,n)}{n}$$
$$= (n-1)!$$
$$P_{\text{ring}}(6,6) = (6-1)!$$
$$= 120$$

**Answer is B.**

**23.** The two possibilities are mutually exclusive. Add the probability of drawing a cat card to the probability of drawing a dog card to find the probability of drawing either.

$$p\{\text{cat,dog}\} = p\{\text{cat}\} + p\{\text{dog}\}$$
$$= \frac{5}{10} + \frac{2}{10}$$
$$= 7/10$$

**Answer is D.**

**24.** The mean is

$$\overline{x} = \frac{\sum x_i}{n} = \frac{17 + 18 + 24 + 33}{4}$$
$$= 23$$

The sample standard deviation is

$$s = \sqrt{\frac{\sum (x_i - \overline{x})^2}{n-1}}$$
$$= \sqrt{\frac{\begin{array}{c}(17-23)^2 + (18-23)^2 \\ + (24-23)^2 + (33-23)^2\end{array}}{4-1}}$$
$$= 7.348 \quad (7.3)$$

**Answer is B.**

**25.** The mean of the Poisson distribution, $\lambda$, is 4 tables. The probability of 7 tables is given by the formula

$$p\{7\} = \frac{e^{-\lambda}\lambda^x}{x!} = \frac{e^{-4}(4)^7}{7!}$$
$$= 0.0595 \quad (0.060)$$

**Answer is A.**

**26.** First calculate the following values.

$$\sum x_i = 2 + 4 + 6 + 7 = 19$$
$$\sum y_i = 10 + 9 + 6 + 4 = 29$$
$$\sum x_i^2 = (2)^2 + (4)^2 + (6)^2 + (7)^2 = 105$$
$$\sum x_i y_i = (2)(10) + (4)(9) + (6)(6) + (7)(4) = 120$$

The slope is

$$b = \frac{\sum x_i y_i - \frac{(\sum x_i)(\sum y_i)}{n}}{\sum x_i^2 - \frac{(\sum x_i)^2}{n}}$$

$$= \frac{120 - \frac{(19)(29)}{4}}{105 - \frac{(19)^2}{4}}$$
$$= -1.203 \quad (-1.2)$$

**Answer is B.**

**27.** Carboxylic acids contain the carboxyl group $-\text{COOH}$.

**Answer is B.**

**28.**  Alkanes are hydrocarbons with the form $C_nH_{2n+2}$.

**Answer is D.**

**29.**  Alkenes are hydrocarbons with the form $C_nH_{2n}$.

**Answer is B.**

**30.**  Ethers are of the form [R]-O-[R].

**Answer is B.**

**31.**  Alkyl halides are alkanes with one hydrogen replaced by a halogen atom.

**Answer is B.**

**32.  Answer is C.**

**33.**
$$2H_2 + O_2 = 2H_2O$$

**Answer is B.**

**34.**  $H_2 + Cl_2 \longrightarrow 2HCl$

Determine the moles of gas in the drum.

$$(10\ kg_{H_2}) \left( \frac{1}{0.002\ \frac{kg}{mol}} \right) = 5000\ mol\ of\ H_2$$

$$(355\ kg_{Cl_2}) \left( \frac{1}{0.071\ \frac{kg}{mol}} \right) = 5000\ mol\ of\ Cl_2$$

There are $(2)(5000\ mol) = 10\,000$ mol of HCl in the drum.

$$p = \frac{n\overline{R}T}{v}$$

$$= \frac{(10\,000\ mol) \left( 8.314\ \frac{N \cdot m}{mol \cdot K} \right) (60°C + 273°)}{0.50\ m^3}$$

$$= 55\,371\,240\ Pa \quad (55.4\ MPa)$$

**Answer is D.**

**35.**  The partial pressure of the $H_2$ is the pressure the $H_2$ would be at if it occupied the entire 1 m³ box.

$$p_{H_2} = \frac{n_{H_2}\overline{R}T}{v}$$

$$n_{H_2} = \frac{m}{MW} = \frac{(2\ kg) \left( 1000\ \frac{g}{kg} \right)}{2.02\ \frac{g}{mol\ H_2}}$$

$$= 990.1\ mol$$

$$p_{H_2} = \frac{n_{H_2}\overline{R}T}{v}$$

$$= \left( \frac{(990.1\ mol) \left( 8.314\ \frac{J}{mol \cdot K} \right) (300K)}{1\ m^3} \right)$$

$$\times \left( \frac{1\ MPa}{10^6\ Pa} \right)$$

$$= 2.47\ MPa \quad (2.5\ MPa)$$

**Answer is A.**

**36.**
$$n_{O_2} = \frac{m}{MW} = \frac{(10\ kg) \left( 1000\ \frac{g}{kg} \right)}{32.0\ \frac{g}{mol\ O_2}}$$

$$= 312.5\ mol$$

$$O_2 + 2H_2 \longrightarrow 2H_2O$$

2 moles of $H_2$ combine with each mole of $O_2$, so

$$(990.1\ mol\ H_2) \left( \frac{1\ mol\ O_2}{2\ mol\ H_2} \right) = 495.1\ mol\ of\ O_2$$

495.1 mol of $O_2$ are needed to burn all the $H_2$. Only 312.5 mol of $O_2$ are available, so there will be excess $H_2$.

moles $H_2O$ produced
$$= (moles\ O_2)(moles\ H_2O\ per\ mole\ O_2)$$
$$= (312.5\ mol) \left( \frac{2\ mol\ H_2O}{mol\ O_2} \right)$$
$$= 625\ mol\ of\ H_2O$$

mass $H_2O = (moles\ H_2O)(molar\ mass)$
$$= (625\ mol\ H_2O) \left( 18.01\ \frac{g}{mol} \right) \left( \frac{1\ kg}{1000\ g} \right)$$
$$= 11.26\ kg \quad (11\ kg)$$

**Answer is B.**

**37.**  The decibel is a logarithmic measure, so the cascaded gains are added.

$$gain_{overall} = 10\ dB + 4\ dB = 14\ dB$$

**Answer is C.**

**38.**
$$\frac{x_o}{x_i} = \frac{G}{1 - GH} = \frac{15}{1 - (15)(0.02)}$$
$$= 21.43 \quad (21)$$

**Answer is D.**

**39.** The system is controllable if $|\mathbf{P}| \neq 0$, where $\mathbf{P} \equiv$ $[\mathbf{B}, \mathbf{AB}]$ for the system $\dot{\mathbf{x}} = \mathbf{Ax} + \mathbf{Bu}$.

(A) $|\mathbf{P}| = \begin{vmatrix} 1 & 9 \\ 2 & 18 \end{vmatrix} = 18 - 18 = 0$

(B) $|\mathbf{P}| = \begin{vmatrix} 10 & 0 \\ 2 & 0 \end{vmatrix} = 0 - 0 = 0$

(C) $|\mathbf{P}| = \begin{vmatrix} 10 & 40 \\ 10 & 40 \end{vmatrix} = 40 - 40 = 0$

(D) $|\mathbf{P}| = \begin{vmatrix} 0 & 12 \\ 4 & 16 \end{vmatrix} = 0 - 48 = -48$

**Answer is D.**

**40.** Taking the Laplace transform of the differential equation,
$$sC(s) = 10R(s) - 16C(s)$$
$$G(s) = \frac{C(s)}{R(s)} = \frac{10}{s + 16}$$

**Answer is B.**

**41.**
$$G(s) = \left( \frac{G_1(s)}{1 + G_1(s)H(s)} \right)(1) \qquad \text{[I]}$$

From Prob. 40,
$$G(s) = \frac{10}{s + 16}$$

Dividing through by $s$,
$$G(s) = \frac{\dfrac{10}{s}}{1 + \dfrac{16}{s}} \qquad \text{[II]}$$

Compare Eqs. I and II.
$$G_1(s) = 10/s$$
$$G_1(s)H(s) = 16/s$$
$$H(s) = \left( \frac{16}{s} \right) \left( \frac{1}{G_1(s)} \right) = \left( \frac{16}{s} \right) \left( \frac{s}{10} \right) = 8/5$$

**Answer is C.**

**42.**
$$\dot{C}(t) = 10r(t) - 16C(t)$$
$$C(0) = 0$$

For $t > 0$,
$$\dot{C}(t) = (10)(2) - 16C(t) = 20 - 16C(t)$$

Take the Laplace transform of both sides.
$$SC(s) = 20 - 16C(s)$$
$$C(s) = \frac{20}{16 + s}$$

The $\lim_{t \to \infty} (C(t))$ is given by the final value theorem.
$$\lim_{t \to \infty} (C(t)) = \lim_{s \to 0} (S(s)) = \lim_{s \to 0} \left[ S \left( \frac{20}{16 + s} \right) \right] = \frac{0}{16}$$
$$= 0$$

**Answer is B.**

**43.** The denominator is zero for input frequencies where the system is unstable.
$$\left. (s^2 - 6s + A) \right|_6 = 0$$
$$A = 36 - 36 = 0$$

**Answer is D.**

**44.**
$$H(s) = \frac{s - B}{s^2 - 6s}$$

Since $H(0)$ does not approach infinity, $B$ must be such that the portion of the denominator that approaches zero as $s$ approaches zero is cancelled by the numerator. Therefore, the numerator must have a root at zero.
$$\left. s - B \right|_{s \to 0} = 0$$
$$B = 0$$

**Answer is A.**

**45.** $H(s)$, with the respective values substituted for $A$ and $B$, becomes
$$H(s) = \frac{s}{s^2 - 6s}$$

Poles are values of $s$ for which $H(s)$ becomes infinite. This occurs where the denominator is zero.
$$s^2 - 6s = s(s - 6) = 0$$
$$s = 0, 6$$
$$H(0) = \frac{0}{0 - 0}$$
$$\lim_{s \to 0} H(s) = \lim_{s \to 0} \left( \frac{1}{2s - 6} \right) = -1/6$$
$$\text{[L'Hôpital's rule]}$$

0 is not a pole.

$$H(6) = \frac{6}{(6)^2 - (6)(6)} = \frac{6}{0}$$

$$\lim_{s \to 6} H(s) \longrightarrow \infty$$

6 is a pole.

**Answer is B.**

**46.** The safety and well-being of society comes before responsibilities to a client or employer.

**Answer is A.**

**47.** Unless an agreement is made regarding patents and inventions before the work is performed, the client who paid for the services should receive rights to all inventions and patents.

**Answer is C.**

**48.** The NCEES sample code of ethics specifically states that registrants may not perform work beyond their level of expertise. The other two examples may be unethical under some circumstances, but are not specifically forbidden by the NCEES Code.

**Answer is C.**

**49.** Each segment must be signed and sealed by the individual registrants responsible for that segment for a registrant to oversee a project.

**Answer is A.**

**50.** Designing a product to fail after the warranty has expired is not an ethical practice. Using the proceeds from one product to reduce the price of another is unethical since it leads to unfair pricing practices and reduced shareholder value. It is ethical, however, to charge a premium for a product that performs a function better or is made of superior materials.

**Answer is B.**

**51.** From the NCEES sample code of ethics, statements may be issued if all interested parties who inspired or paid for the announcement are explicitly identified, even if compensation is received for the statements. Of course, statements must not be biased.

**Answer is D.**

**52.** Professionals have ethical responsibilities to their employers, consumers, and competitors.

**Answer is D.**

**53.** It is unethical to indiscriminately criticize other registrants or their work, even if truthful, unless the public's safety or welfare would be compromised without the criticism.

**Answer is B.**

**54.** Registration laws are passed by state legislatures. Professional societies and the national board of registration lobby the state legislatures to ensure the requirements for registration are kept current.

**Answer is A.**

**55.**
$$R = 1.5P$$
$$C = 25{,}000 + 0.03P$$
$$R = C$$
$$1.5P = 25{,}000 + 0.03P$$
$$P = \frac{25{,}000}{1.47}$$
$$= 17{,}007 \quad (17{,}000)$$

**Answer is C.**

**56.**
$$C = (0.8)(5000 \text{ units})\left(\frac{\$5}{\text{unit}}\right) + \$3000$$
$$= \$23{,}000$$
$$R = (0.8)(5000 \text{ units})\left(\frac{\$16}{\text{unit}}\right)$$
$$= \$64{,}000$$
$$P = R - C$$
$$= \$41{,}000$$

**Answer is A.**

**57.**

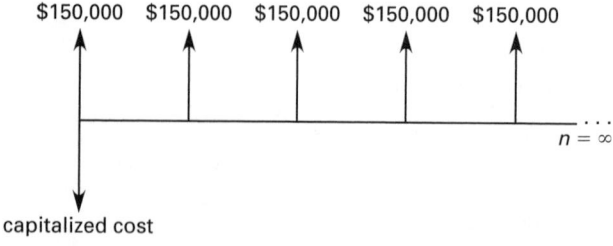

An equivalent annual payment $(A)$ based on the $150,000 disbursement is

$$A = P(A/P, i\%, n)$$
$$= (\$150,000)(A/P, 8\%, 15)$$
$$= (\$150,000)(0.1168)$$
$$= \$17,520$$
$$\text{capitalized cost} = \frac{\$17,520}{0.08}$$
$$= \$219,000$$

**Answer is C.**

**58.** 10% is a nominal rate, not an effective rate, because the compounding is done more than once per year. The effective annual rate is

$$i = \left(1 + \frac{0.10}{2}\right)^2 - 1$$
$$= 0.1025$$
$$P = \frac{\$1000}{(1 + 0.1025)^6}$$
$$= (\$1000)(0.5568)$$
$$= \$556.80 \quad (\$557)$$

**Answer is B.**

**59.**
$$1 + i = (1 + \phi)^k$$
$$\phi = (1 + i)^{\frac{1}{k}} - 1$$
$$= (1 + 0.12)^{\frac{1}{12}} - 1$$
$$= 0.00949 \quad \text{[effective rate per month]}$$
$$\$500 = P(1 + \phi)^{11}$$
$$P = \frac{\$500}{(1 + 0.00949)^{11}}$$
$$= \$450.7 \quad (\$451)$$

**Answer is C.**

**60.** Calculate the simple interest by multiplying the present worth, $P$, by the interest rate per period, $i$, by the number of compounding periods, $n$.

$$Pin = (\$5000)(0.08)(5)$$
$$= \$2000$$

**Answer is B.**

**61.** First calculate the effective interest rate per period.

$$i = \frac{r}{m} = \frac{0.06}{12}$$
$$= 0.005 \quad (0.5\%)$$

Use the uniform series sinking fund equation where the future worth, $F$, is $1000, the interest rate, $i$, is 0.5%, and the number of compounding periods, $n$, is 12.

$$A = F(A/F, i\%, n)$$
$$= F\left(\frac{i}{(1 + i)^n - 1}\right)$$
$$= (\$1000)\left(\frac{0.005}{(1 + 0.005)^{12} - 1}\right)$$
$$= \$81.07 \quad (\$81)$$

**Answer is C.**

**62.** Use the cash flow factor $(A/P, i\%, n)$, where the interest rate, $i$, is 7% and the number of periods, $n$, is 10.

$$A = P(A/P, i\%, n)$$
$$= P\left(\frac{i(1 + i)^n}{(1 + i)^n - 1}\right)$$
$$= (\$9000)\left(\frac{(0.07)(1.07)^{10}}{(1.07)^{10} - 1}\right)$$
$$= \$1281.40 \quad (\$1300)$$

**Answer is D.**

**63.** The irregular cash flow can be divided into two standard cash flows. The first three years form a uniform annual payment cash flow. An equivalent single payment at the end of the third year can be calculated, and then interest for the last two years can be treated as a single payment cash flow.

Work backward from the known final sum. First consider only the second cash flow—the interest. Calculate a single payment at the end of three years that will give a future value of $60,000 at the end of five years. Use the cash flow factor $(P/F, i\%, n)$, where the interest rate, $i$, is 8% and the number of periods, $n$, is 2. The value of $(P/F, 8\%, 2)$ is found in a table of standard cash flow factors.

$$P_{4 \text{ yr}} = F(P/F, 8\%, 2)$$
$$= (\$60,000)(0.8573)$$
$$= \$51,438$$

This amount, $51,438, then becomes the future value for the three uniform annual payments. In order to have $51,438 at the end of three years, the amount that must be paid into the fund at the end of each year is

$$A = F(A/F, 8\%, 3)$$
$$= (\$51,438)(0.3080)$$
$$= \$15,842.90 \quad (\$16,000)$$

**Answer is A.**

**64.**

$T_1$ = tension in left cord

$T_2$ = tension in right cord

$2T_2 = 1000$ N

$\sum F = 0$   [pulley 3]

$2T_1 = T_2$

$\sum F = 0$   [pulley 2]

$$P = T_1 = \frac{T_2}{2}$$

$$= \frac{500 \text{ N}}{2}$$

$$= 250 \text{ N}$$

**Answer is B.**

**65.**

$F_f$ = frictional force = $\mu N = \mu W$

For sliding conditions,

$$\sum F_x = P_{\text{slide}} - F_f = 0$$

$$P_{\text{slide}} = F_f = \mu W$$

For tipping conditions,

$$+\sum M_O = P_{\text{tip}}a - W\left(\frac{a}{2}\right) = 0$$

$$P_{\text{tip}} = \frac{W}{2}$$

When tipping and sliding conditions are equal,

$$P_{\text{tip}} = P_{\text{slide}}$$

$$\frac{W}{2} = \mu W$$

$$\mu = 0.5$$

**Answer is D.**

**66.**

$$F_x = (500 \text{ N})\left(\frac{7}{\sqrt{(3)^2 + (7)^2 + (8)^2}}\right)$$

$$= 316.8 \text{ N}$$

$$F_y = (500 \text{ N})\left(\frac{3}{\sqrt{(3)^2 + (7)^2 + (8)^2}}\right)$$

$$= 135.8 \text{ N}$$

$$F_z = (500 \text{ N})\left(\frac{8}{\sqrt{(3)^2 + (7)^2 + (8)^2}}\right)$$

$$= 362.1 \text{ N}$$

**Answer is B.**

**67.** From the impulse-momentum principle,

$$F\Delta t = m\Delta v$$

**Answer is C.**

**68.** Disregarding air friction,

$$H = \frac{v_0^2 \sin^2 \phi}{2g} = \frac{\left(1250 \frac{\text{m}}{\text{s}}\right)^2 (\sin^2 30°)}{(2)\left(9.81 \frac{\text{m}}{\text{s}^2}\right)\left(1000 \frac{\text{m}}{\text{km}}\right)}$$

$$= 19.9 \text{ km}   (20 \text{ km})$$

**Answer is A.**

**69.**
$$R = \frac{v_0^2 \sin 2\phi}{g} = \frac{\left(1250 \frac{\text{m}}{\text{s}}\right)^2 \sin((2)(30°))}{9.81 \frac{\text{m}}{\text{s}^2}}$$

$$= 137\,938 \text{ m}   (140 \text{ km})$$

**Answer is C.**

**70.**
$$v = \sqrt{2gh} = \sqrt{(2)\left(9.81 \frac{\text{m}}{\text{s}^2}\right)(48 \text{ m})}$$

$$= 30.7 \text{ m/s}   (31 \text{ m/s})$$

**Answer is D.**

**71.**
$$0.9 = \frac{v'}{v}$$

$$v' = 0.9v$$

$$E_k = \frac{m(v')^2}{2}$$

$$= \frac{(60 \text{ kg})\left((0.9)\left(30.7 \frac{\text{m}}{\text{s}}\right)\right)^2}{2}$$

$$= 22\,900 \text{ J}   (23 \text{ kJ})$$

**Answer is C.**

**72.**
$$\sum M_O = 0$$

$$-200 \text{ N·m} - (100 \text{ N})\left(\tfrac{3}{5}\right)(2.5 \text{ m})$$
$$- (2.5 \text{ m})(300 \text{ N}) + R(500 \text{ N}) = 0$$
$$R = 2.2 \text{ m}$$

**Answer is D.**

**73.**
$$- (2.5 \text{ m})(300 \text{ N}) - (2.5 \text{ m})(100 \text{ N})\left(\tfrac{3}{5}\right)$$
$$+ (1.0 \text{ m})F_3 = 0$$
$$F_3 = 900 \text{ N}$$

**Answer is D.**

**74.**

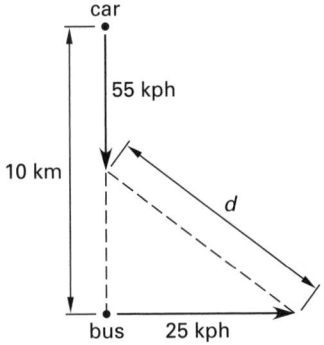

The distance, $d$, between the car and bus for time, $t$, is

$$d = \sqrt{(10 - 55t)^2 + (25t)^2}$$
$$= \sqrt{3650t^2 - 1100t + 100}$$

**Answer is D.**

**75.**
$$v_{\text{rel}} = \sqrt{(-55 \text{ kph})^2 + (25 \text{ kph})^2}$$
$$= 60.4 \text{ kph} \quad (60 \text{ kph})$$

**Answer is D.**

**76.**
$$\tau = \frac{VQ}{Ib} = \frac{V\left(\frac{1}{2}bh\right)\left(\frac{h}{4}\right)}{\left(\frac{bh^3}{12}\right)b}$$

$$\tau_{\text{max}} = \left(\frac{3}{2}\right)\left(\frac{V}{bh}\right) = \frac{3V}{2A}$$
$$= \frac{(3)(7000 \text{ N})}{(2)(4 \text{ cm})(6 \text{ cm})}$$
$$= 437.5 \text{ N/cm}^2 \quad (440 \text{ N/cm}^2)$$

**Answer is B.**

**77.**
$$J = \frac{\pi r^4}{2} = \frac{\pi \left(\frac{0.250 \text{ cm}}{2}\right)^4}{2}$$
$$= 3.83 \times 10^{-4} \text{ cm}^4$$

$$T = \frac{JG\phi}{L}$$

$$= \frac{(3.83 \times 10^{-4} \text{ cm}^4)\left(11.5 \times 10^6 \, \dfrac{\text{N}}{\text{cm}^2}\right)}{6 \text{ cm}}$$
$$\times (8°)\left(\frac{2\pi \text{ rad}}{360°}\right)$$

$$= 102.6 \text{ N·cm} \quad (100 \text{ N·cm})$$

**Answer is A.**

**78.** Regardless of a cantilever beam's loading, the deflection equations contain $I$ in the denominator. Base and height dimensions do not appear anywhere else.

$$I = \frac{bh^3}{3}$$

$$\frac{y_{0.5}}{y_{0.25}} = \frac{I_{0.25}}{I_{0.5}} = \frac{(0.25 \text{ cm})^4}{(0.5 \text{ cm})^4}$$
$$= 0.0625$$

**Answer is A.**

**79.** $\delta_{\text{total}} = \delta_{\text{aluminum}} + \delta_{\text{steel}}$

$$= \left(\frac{FL}{AE}\right)_{\text{aluminum}} + \left(\frac{FL}{AE}\right)_{\text{steel}}$$

$$= \frac{(40\,000 \text{ N})(4 \text{ cm})}{\left(\dfrac{\pi(1.7 \text{ cm})^2}{4}\right)\left(6.9 \times 10^6 \, \dfrac{\text{N}}{\text{cm}^2}\right)}$$

$$+ \frac{(40\,000 \text{ N})(4 \text{ cm})}{\left(\dfrac{\pi(0.8 \text{ cm})^2}{4}\right)\left(20.7 \times 10^6 \, \dfrac{\text{N}}{\text{cm}^2}\right)}$$

$$= 0.0256 \text{ cm} \quad (0.026 \text{ cm})$$

**Answer is C.**

**80.** $\sigma = \sigma_{\text{axial}} + \sigma_{\text{bending}}$

$$= \frac{F}{A} + \frac{Mc}{I} = \frac{F}{bh} + \frac{Fd\left(\dfrac{h}{2}\right)}{\frac{1}{12}bh^3}$$

$$= \frac{1500 \text{ N}}{(1.5 \text{ cm})(0.75 \text{ cm})}$$

$$+ \frac{(1500 \text{ N})(3 \text{ cm})\left(\dfrac{0.75 \text{ cm}}{2}\right)}{\left(\dfrac{1}{12}\right)(1.5 \text{ cm})(0.75 \text{ cm})^3}$$

$$= 33\,333 \text{ N/cm}^2 \quad (33.3 \text{ kN/cm}^2)$$

**Answer is D.**

**81.**
$$n = \frac{20 \times 10^6 \, \dfrac{\text{N}}{\text{cm}^2}}{1 \times 10^6 \, \dfrac{\text{N}}{\text{cm}^2}} = 20$$

$$\text{transformed steel width} = (\text{steel width})n$$
$$= (2 \text{ cm})(20)$$
$$= 40 \text{ cm}$$

$$y_c = \frac{\begin{array}{c}(2\text{ cm})(2\text{ cm})(1.125\text{ cm}) \\ + (0.125\text{ cm})(40\text{ cm})(0.0625\text{ cm})\end{array}}{4\text{ cm} + 5\text{ cm}}$$
$$= 0.535\text{ cm}\quad(0.54\text{ cm})$$

**Answer is B.**

**82.**
$$\Delta d = \epsilon_{\text{elastic limit}}\nu d_1$$
$$= (0.015)(0.3)(0.25\text{ cm})$$
$$= 0.001125\text{ cm}\quad(11\ \mu\text{m})$$

**Answer is A.**

**83.**
$$y = \text{deflection}$$
$$y' = \frac{dy}{dx} = \text{slope}$$
$$y'' = \frac{d^2y}{dx^2} = \frac{M(x)}{EI}\quad[\text{moment}]$$

**Answer is B.**

**84.** Use Euler's formula, where $K = 2$ for one fixed end.

$$P_{\text{cr}} = \frac{\pi^2 EI}{(KL)^2}$$
$$= \frac{\pi^2(200 \times 10^9\text{ Pa})(1.0 \times 10^{-4}\text{ m}^4)}{((2)(3.5\text{ m}))^2}$$
$$= 4\,028\,409\text{ N}$$
$$P_{\text{allowable}} = \frac{P_{\text{cr}}}{\text{FS}} = \frac{4\,028\,409\text{ N}}{3.0}$$
$$= 1\,342\,803\text{ N}\quad(1343\text{ kN})$$

**Answer is C.**

**85.  Answer is A.**

**86.** The cooling rate affects the hardness but not the hardenability.

**Answer is B.**

**87.  Answer is D.**

**88.** A standard tensile test can determine the elastic modulus, yield strength, ultimate tensile strength, and ductility of a metal.

**Answer is D.**

**89.** The region where strain hardening occurs is between points B and C. O to A is known as the linear region, A to B is the plastic region, and C to D is where reduction occurs.

**Answer is C.**

**90.** Gibb's phase rule is generally expressed as $P+F = C + 2$. $P$ is the number of phases existing simultaneously, $F$ is the number of independent variables (known as degrees of freedom), and $C$ is the number of elements in the alloy. Composition, temperature, and pressure are examples of degrees of freedom that can be varied.

**Answer is C.**

**91.** From Hooke's law,

$$\varepsilon = \frac{\sigma}{E} = \frac{\delta}{L}$$
$$\delta = \frac{\sigma L}{E} = \frac{\left(120 \times 10^6\ \dfrac{\text{N}}{\text{m}^2}\right)(3\text{ m})}{70 \times 10^9\ \dfrac{\text{N}}{\text{m}^2}}$$
$$= 5.14 \times 10^{-3}\text{ m}\quad(5.1\text{ mm})$$

**Answer is B.**

**92.** The Poisson ratio must be in the range $0 < \nu < 0.5$. Option B is the only answer that satisfies this condition.

**Answer is B.**

**93.** The variable $D$ is the pipe diameter.

**Answer is B.**

**94.**
$$p = \rho g h$$
$$h = \frac{p}{\rho g} = \frac{p}{(\text{SG})\rho_{\text{water}}g}$$
$$= \frac{\left(1000\ \dfrac{\text{N}}{\text{m}^2}\right)\left(100\ \dfrac{\text{cm}}{\text{m}}\right)}{(1.56)\left(1000\ \dfrac{\text{kg}}{\text{m}^3}\right)\left(9.81\ \dfrac{\text{m}}{\text{s}^2}\right)}$$
$$= 6.53\text{ cm}\quad(6.5\text{ cm})$$

**Answer is B.**

**95.** Inertial and gravitational forces dominate for a spillway. The Froude numbers must be equal.

$$(N_{\text{Fr}})_{\text{dam}} = (N_{\text{Fr}})_{\text{model}}$$

$$\left(\frac{v}{\sqrt{gL}}\right)_{\text{dam}} = \left(\frac{v}{\sqrt{gL}}\right)_{\text{model}}$$

$$v_{\text{dam}} = v_{\text{model}} \sqrt{\frac{L_{\text{dam}}}{L_{\text{model}}}} = \left(5 \ \frac{\text{m}}{\text{s}}\right) \sqrt{\frac{15}{1}}$$

$$= 19.36 \ \text{m/s} \quad (19 \ \text{m/s})$$

**Answer is D.**

**96.   Answer is D.**

**97.** Apply a small tilt.

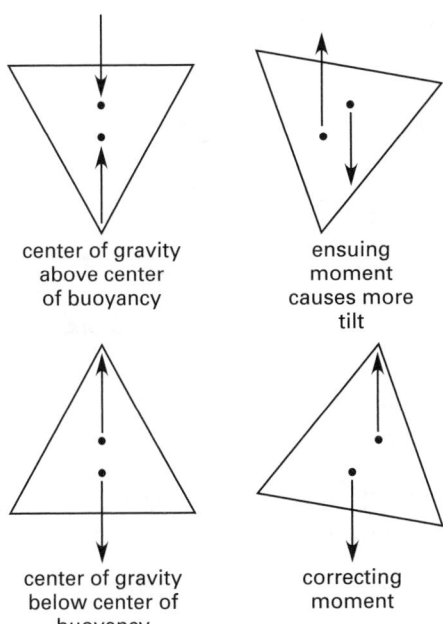

center of gravity
above center
of buoyancy

ensuing
moment
causes more
tilt

center of gravity
below center of
buoyancy

correcting
moment

**Answer is B.**

**98.** $\quad v = \sqrt{2gh} = \sqrt{(2)\left(9.81 \ \dfrac{\text{m}}{\text{s}^2}\right)(10 \ \text{m})}$

$$= 14 \ \text{m/s}$$

**Answer is C.**

**99.   Answer is D.**

**100.** $h_{\text{air}} = (h_{\text{water}})\left(\dfrac{\rho_{\text{water}}}{\rho_{\text{air}}}\right)$

$$= \frac{(6 \ \text{cm water})\left(\dfrac{1000 \ \dfrac{\text{kg}}{\text{m}^3}}{1.2 \ \dfrac{\text{kg}}{\text{m}^3}}\right)}{100 \ \dfrac{\text{cm}}{\text{m}}}$$

$$= 50 \ \text{m of air}$$

$$P = hQ\rho g$$

$$= \frac{(50 \ \text{m})\left(110 \ \dfrac{\text{m}^3}{\text{min}}\right)\left(1.2 \ \dfrac{\text{kg}}{\text{m}^3}\right)\left(9.81 \ \dfrac{\text{m}}{\text{s}^2}\right)}{\left(60 \ \dfrac{\text{s}}{\text{min}}\right)}$$

$$= 1079 \ \text{W} \quad (1.1 \ \text{kW})$$

**Answer is C.**

**101.** The hydraulic radius is found by dividing the area of the channel by the wetted perimeter. The wetted perimeter of the channel is found using the following equation, where $y$ is the depth and $m$ is the side slope.

$$P = 2y\sqrt{1 + m^2}$$

$$= (2)(5 \ \text{m})\sqrt{1 + (1)^2}$$

$$= 14.14 \ \text{m}$$

Find the area using the following equation.

$$A = my^2$$

$$= (1)(5 \ \text{m})^2$$

$$= 25 \ \text{m}^2$$

Calculate the hydraulic radius.

$$R = \frac{A}{P} = \frac{25 \ \text{m}^2}{14.14 \ \text{m}}$$

$$= 1.77 \ \text{m} \quad (1.8 \ \text{m})$$

**Answer is B.**

**102.   Answer is C.**

**103.** This is a current divider.

$$i_{6 \, \Omega} = (6 \ \text{A})\left(\frac{18 \ \Omega}{6 \ \Omega + 18 \ \Omega}\right)$$

$$= 4.5 \ \text{A}$$

**Answer is D.**

**104.**
$$C_{\text{eq}} = 30\ \mu\text{F} + \cfrac{1}{\cfrac{1}{20\ \mu\text{F}} + \cfrac{1}{20\ \mu\text{F}}}$$
$$= 40\ \mu\text{F}$$

Answer is C.

**105.**    $f_o = \cfrac{1}{2\pi\sqrt{LC}} = \cfrac{1}{2\pi\sqrt{(4\ \text{H})(3\times 10^{-6}\ \text{F})}}$
$$= 45.94\ \text{Hz}\quad(46\ \text{Hz})$$

Answer is C.

**106.**

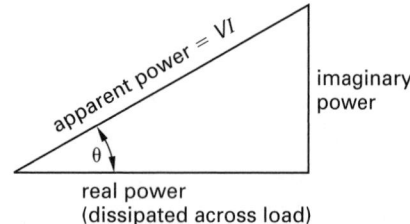

$$\text{power factor} = \cos\theta = \frac{\text{real power}}{\text{apparent power}}$$

Answer is B.

**107.**
$$V_{\text{line}} = \sqrt{3}V_{\text{phase}}$$
$$V_{\text{ab}} = \sqrt{3}(110\ \text{V})$$
$$= 190.5\ \text{V}\quad(190\ \text{V})$$

Answer is C.

**108.**
$$W = Fd = \frac{Vqd}{r}$$
$$\frac{V}{r} = 50\ \text{V/m}$$
$$q = 10\ \text{C}$$
$$d = 5\ \text{m}$$
$$W = \left(50\ \frac{\text{W}}{\text{m}}\right)(10\ \text{C})(5\ \text{m})$$
$$= 2500\ \text{W·C}$$
$$= 2500\ \text{J}\quad(2.5\ \text{kJ})$$

Answer is C.

**109.**
$$\omega_o = \frac{1}{\sqrt{LC}}$$
$$\alpha = \frac{1}{2RC}$$
$$\text{overdamped if } \alpha > \omega_o$$
$$\text{underdamped if } \omega_o > \alpha$$

Answer is D.

**110.**
$$R_{\text{AB}} = \left(\left(50\ \Omega + \left(\frac{1}{40\ \Omega} + \frac{1}{40\ \Omega} + \frac{1}{50\ \Omega}\right)^{-1}\right)^{-1} + \frac{1}{30\ \Omega}\right)^{-1}$$
$$= 20.45\ \Omega$$

Answer is B.

**111.**    $V_C(t) = V_B\left(1 - e^{-\frac{t}{RC}}\right)$
$$\frac{V_C(t)}{V_B} = 0.80 = 1 - e^{-\frac{t}{(150)(100\times 10^{-6})}}$$
$$= e^{-\frac{t}{0.015\ \text{s}}}$$
$$= 0.20\ \text{s}$$
$$t = (0.015)\left(\ln(0.20\ \text{s})\right)$$
$$= 0.02414\ \text{s}\quad(24\ \text{ms})$$

Answer is D.

**112.**  Answer is C.

**113.**

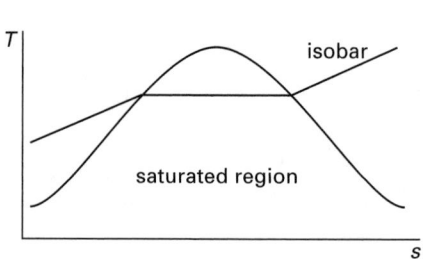

Answer is C.

**114.**  For a constant pressure process,
$$W = mR(T_2 - T_1)$$
$$= \frac{(1\ \text{kg})\left(278\ \dfrac{\text{J}}{\text{kg·K}}\right)(600\text{K} - 500\text{K})}{1000\ \dfrac{\text{J}}{\text{kJ}}}$$
$$= 27.8\ \text{kJ}\quad(58\ \text{kJ})$$

Answer is B.

**115.**    $v = (1-x)v_{f,T_{\text{sat}}=278\text{K}} + xv_{g,T_{\text{sat}}=278\text{K}}$
$$= (1 - 0.1)\left(0.001\,58\ \frac{\text{m}^3}{\text{kg}}\right)$$
$$+ (0.1)\left(0.2479\ \frac{\text{m}^3}{\text{kg}}\right)$$
$$= 0.0262\ \text{m}^3/\text{kg}\quad(0.026\ \text{m}^3/\text{kg})$$

Answer is B.

**116.**
$$Q = m(h_{\text{final}} - h_{\text{initial}})$$

From the saturated water table and the superheated water table,

$$Q = (1 \text{ kg}) \left( 3704 \, \frac{\text{kJ}}{\text{kg}} - 419 \, \frac{\text{kJ}}{\text{kg}} \right)$$
$$= 3285 \text{ kJ} \quad (3.3 \text{ MJ})$$

**Answer is A.**

**117.** Vaporization and condensation are isobaric processes.

**Answer is D.**

**118.**
$$Q = V \rho c_p (T_{\text{exit}} - T_{\text{entrance}})$$
$$T_{\text{entrance}} = T_{\text{exit}} - \frac{Q}{V \rho c_p}$$
$$= 48°\text{C}$$
$$- \frac{26\,400 \times 10^6 \text{ J}}{(210 \text{ m}^3) \left( 1000 \, \frac{\text{kg}}{\text{m}^3} \right) \left( 4187 \, \frac{\text{J}}{\text{kg·}°\text{C}} \right)}$$
$$= 18°\text{C}$$

**Answer is B.**

**119.** An extensive property depends on the amount of mass present. Internal energy has this trait.

**Answer is D.**

**120.** **Answer is C.**

# Solutions for Exam 2–Afternoon Section

**1.** The vertex is at $(0, 8)$, so the equation is

$$(x - h)^2 = 4p(y - k)$$
$$(x - 0)^2 = 4p(y - 8)$$

Substituting the known point $(3, 0)$,

$$x^2 = 4p(y - 8)$$
$$(3)^2 = 4p(0 - 8)$$
$$9 = -32p$$
$$p = -9/32$$

The complete equation of the parabola is

$$x^2 = (4)\left(\frac{-9}{32}\right)(y - 8) = \frac{-9y}{8} + 9$$

Rearranging and taking the derivative,

$$2x + \left(\frac{9}{8}\right)\left(\frac{dy}{dx}\right) = 0$$
$$m = \text{slope} = \frac{dy}{dx} = \frac{-16x}{9}$$
$$\left.\frac{dy}{dx}\right|_{x=2} = -\frac{32}{9}$$

The equation is

$$y = \frac{-32}{9}x + b$$

Determine the $y$-intercept by substituting the known point.

$$\frac{40}{9} = \left(\frac{-32}{9}\right)(2) + b$$
$$b = \frac{104}{9}$$
$$y = -\frac{32}{9}x + \frac{104}{9}$$
$$x = -\frac{9}{32}y + \frac{13}{4}$$

**Answer is C.**

**2.**
$$m = \text{mass of C}^{14} = A_0 e^{-\lambda t}$$
$$\frac{1}{2} = e^{-\lambda(5730 \text{ yr})}$$
$$\lambda = \frac{-\ln\frac{1}{2}}{5730 \text{ yr}}$$
$$= 1.21 \times 10^{-4}/\text{yr} \quad (1.2 \times 10^{-4}/\text{yr})$$

**Answer is B.**

**3.**
$$\frac{m}{m_0} = e^{-\left(1.21 \times 10^{-4}\,\frac{1}{\text{yr}}\right)(10{,}000 \text{ yr})}$$
$$= e^{-1.21}$$
$$= 0.298$$
$$\text{percent decayed} = 1 - 0.298$$
$$= 0.702 \quad (70\%)$$

**Answer is C.**

**4.** For $y = x^3$, $n = 3$.
$$\text{area} = \frac{bh}{n + 1} = \frac{(3)(27)}{3 + 1} = \frac{81}{4}$$
$$\bar{x} = \left(\frac{n + 1}{n + 2}\right)b = \left(\frac{4}{5}\right)(3) = 2.4$$
$$\bar{y} = \left(\frac{h}{2}\right)\left(\frac{n + 1}{2n + 1}\right) = \left(\frac{27}{2}\right)\left(\frac{4}{7}\right)$$
$$= 7.7$$

**Answer is D.**

**5.**
$$m = \frac{dy}{dx} = 9x^2 - 6x - 24$$
$$m_{x=-3} = (9)(-3)^2 - (6)(-3) - 24 = 75$$

**Answer is C.**

**6.**
$$\frac{dy}{dx} = 0$$
$$9x^2 - 6x - 24 = 0$$
$$x = \frac{6}{18} \pm \frac{\sqrt{36 - (4)(9)(-24)}}{18} = \frac{1}{3} \pm \frac{5}{3}$$
$$= 2, -\frac{4}{3}$$
$$y_{x=2} = (3)(2)^3 - (3)(2)^2 - (24)(2) + 25$$
$$= -11$$

$$y_{x=-\frac{4}{3}} = (3)\left(-\frac{4}{3}\right)^3 - (3)\left(-\frac{4}{3}\right)^2$$
$$- (24)\left(-\frac{4}{3}\right) + 25$$
$$= 44.6$$
$$(2, -11)$$

**Answer is B.**

7.   First calculate the following values.

$$\sum x_i = 8 + 2 + (-3) + (-9) = -2$$

$$\sum y_i = -1 + 2 + 6 + 11 = 18$$

$$\sum x_i^2 = (8)^2 + (2)^2 + (-3)^2 + (-9)^2 = 158$$

$$\sum y_i^2 = (-1)^2 + (2)^2 + (6)^2 + (11)^2 = 162$$

$$\sum x_i y_i = (8)(-1) + (2)(2) + (-3)(6) + (-9)(11)$$
$$= -121$$

The correlation coefficient is given by

$$r = \frac{\sum x_i y_i - \frac{\left(\sum x_i\right)\left(\sum y_i\right)}{n}}{\sqrt{\left(\sum x_i^2 - \frac{\left(\sum x_i\right)^2}{n}\right)\left(\sum y_i^2 - \frac{\left(\sum y_i\right)^2}{n}\right)}}$$

$$= \frac{-121 - \frac{(-2)(18)}{4}}{\sqrt{\left(158 - \frac{(-2)^2}{4}\right)\left(162 - \frac{(18)^2}{4}\right)}}$$

$$= -0.9932 \quad (-0.99)$$

**Answer is D.**

8.   This is a problem in binomial distribution.

$$p = 0.3$$
$$q = 1 - p = 1 - 0.3$$
$$= 0.7$$
$$p\{3\} = \left(\frac{n!}{x!(n-x)!}\right)p^x q^{n-x}$$
$$= \left(\frac{10!}{(3!)(10-3)!}\right)(0.3)^3(0.7)^{10-3}$$
$$= 0.2668 \quad (0.27)$$

**Answer is B.**

9.   Obtain a linear relationship by letting $z$ equal $\sqrt{x}$. Then plot a straight line of the form $y = a + bz$ through the points (1.414,8), (2.236,11), (3.162,15), and (4.583, 20).

Calculate the following values.

$$\sum z_i = 1.414 + 2.236 + 3.162 + 4.583 = 11.395$$

$$\sum y_i = 8 + 11 + 15 + 20 = 54$$

$$\sum z_i^2 = (1.414)^2 + (2.236)^2 + (3.162)^2 + (4.583)^2$$
$$= 38$$

$$\sum z_i y_i = (1.414)(8) + (2.236)(11) + (3.162)(15)$$
$$+ (4.583)(20)$$
$$= 174.998$$

The slope of the line $y = a + bz$ can be found by

$$b = \frac{\sum z_i y_i - \frac{\left(\sum z_i\right)\left(\sum y_i\right)}{n}}{\sum z_i^2 - \frac{\left(\sum z_i\right)^2}{n}}$$

$$= \frac{174.998 - \frac{(11.395)(54)}{4}}{38 - \frac{(11.395)^2}{4}}$$

$$= 3.822 \quad (3.8)$$

**Answer is A.**

10.   Because a specific direction in the variation is not given, a two-tail hypothesis test is used. The $z$-value corresponding to the confidence level is

$$z = \left|\frac{\bar{x} - \mu}{\frac{\sigma}{\sqrt{n}}}\right| = \left|\frac{135\,\frac{\text{kg}}{\text{d}} - 144\,\frac{\text{kg}}{\text{d}}}{\frac{25\,\frac{\text{kg}}{\text{d}}}{\sqrt{30}}}\right|$$

$$= 1.97$$

Comparing this value with a table of $z$-values for various confidence levels shows that the confidence interval is most nearly 95%.

**Answer is C.**

11.   The probability of not having failed before time $t$ is the reliability.

$$\lambda = \frac{1}{\text{MTTF}} = \frac{1}{650\text{ h}}$$
$$= 0.00154\text{ h}^{-1}$$
$$R\{750\text{ h}\} = e^{-\lambda t}$$
$$= e^{(-0.00154\text{ h}^{-1})(750\text{ h})}$$
$$= 0.315 \quad (0.32)$$

**Answer is D.**

**12.** Determine the estimated exposure dose (EED) for the toxicants.

$$\text{EED}_{\text{CuCN}} = \frac{(E_{\text{CuCN}})(\text{CR})}{m} = \frac{\left(40\ \frac{\mu\text{g}}{\text{L}}\right)\left(2\ \frac{\text{L}}{\text{d}}\right)}{70\ \text{kg}}$$

$$= 1.14\ \mu\text{g/kg·d}$$

Use the EED and the reference dose to determine the hazard ratio.

$$\text{HR} = \frac{\text{EED}}{\text{RfD}}$$

$$\text{HR}_{\text{CuCN}} = \frac{\text{EED}_{\text{CuCN}}}{\text{RfD}_{\text{CuCN}}} = \frac{1.14\ \frac{\mu\text{g}}{\text{kg·d}}}{5\ \frac{\mu\text{g}}{\text{kg·d}}}$$

$$= 0.228$$

Similarly, for the other toxicants,

| toxicant | exposure, $E$ ($\mu$g/L) | estimated exposure dose, EED ($\mu$g/kg·d) | RfD ($\mu$g/kg·d) | hazard ratio, HR |
|---|---|---|---|---|
| CuCN | 40 | 1.14 | 5 | 0.228 |
| CH$_3$OH | 1000 | 28.57 | 500 | 0.057 |
| KCN | 600 | 17.14 | 50 | 0.343 |
| | | | total | 0.628 |

**Answer is C.**

**13.** The following results pertain for the dust concentrations and percents of free silica given.

$$\text{PEL}_{0800} = \frac{10\ \frac{\text{mg}}{\text{m}^3}}{5.8 + 2}$$

$$= 1.28\ \text{mg/m}^3$$

The following table is developed similarly.

| percent SiO$_2$ | PEL (mg/m$^3$) |
|---|---|
| 5.8 | 1.28 |
| 6.3 | 1.20 |
| 4.2 | 1.61 |
| 3.9 | 1.69 |
| 4.1 | 1.64 |

The dust concentrations at 0800, 1300, and 1700 hours exceeded the PEL.

**Answer is D.**

**14.** Passive diffusion is influenced by lipid solubility, molecular size, and degree of ionization. The participation of a carrier molecule is required for facilitated diffusion, not passive diffusion.

**Answer is D.**

**15.**
$$(A/P, 7\%, 15) = \frac{i(1+i)^n}{(1+i)^n - 1}$$

$$= \frac{(0.07)(1.07)^{15}}{(1.07)^{15} - 1}$$

$$= 0.1098$$

$$A = P\,(A/P, 7\%, 15)$$

$$= (\$25{,}000)(0.1098)$$

$$= \$2745$$

**Answer is C.**

**16.** The initial principal balance is \$25,000. The interest on this amount is

$$(0.07)(\$25{,}000) = \$1750$$

The amount paid toward principal is

$$\$2745 - \$1750 = \$995$$

The fraction of the payment that is principal is

$$\frac{\$995}{\$2745} = 0.362 \quad (36\%)$$

**Answer is B.**

**17.** Work with the actual cash flows. At $t = 0$, the investor receives \$25,000 from the bank and invests the entire amount in the stock market, a net change of zero. Each year, the investor pays \$2745. At $t = 15$, the investor withdraws the accumulated stock worth $(\$25{,}000)(F/P, 10\%, 15)$.

The present worth of this transaction is

$$P = \$25{,}000 - \$25{,}000 - (\$2745)\,(P/A, 10\%, 15)$$
$$+ (\$25{,}000)\,(F/P, 10\%, 15)\,(P/F, 10\%, 15)$$
$$= \$25{,}000 - \$25{,}000 - (\$2745)(7.6061) + \$25{,}000$$
$$= \$4121$$

**Answer is B.**

**18.** The cost of owning and operating the machine one more year is equal to the operating costs plus the lost salvage value. After 4 yr, the machine's salvage value is $24,000, and after another year it will be $20,000. The lost salvage value is not just $24,000 minus $20,000, however, because the value of having the cash one year earlier must also be considered. The future value of $24,000 after 1 yr is

$$F = (\$24{,}000)\,(F/P, 12\%, 1)$$
$$= (\$24{,}000)(1.1200)$$
$$= \$26{,}880$$

The salvage value lost by waiting a year is $26,880 minus $20,000, or $6880. The cost of one more year of ownership and operation is

$$C = \text{operating cost} + \text{lost salvage value}$$
$$= \$3500 + \$6880$$
$$= \$10{,}380 \quad (\$10{,}000)$$

**Answer is C.**

**19.** The straight line depreciation value is equal to the initial cost minus the salvage value, divided by the life.

$$D = \frac{C - S_n}{n} = \frac{\$23{,}000 - \$1800}{10}$$
$$= \$2120 \quad (\$2100)$$

**Answer is D.**

**20.** The equivalent uniform annual cost (EUAC) is the sum of the equivalent annual values of its initial and maintenance costs, minus the equivalent annual value of its salvage value.

$$\text{EUAC} = (\$15{,}000)\,(A/P, 6\%, 17) + \$550$$
$$- (\$3000)\,(A/F, 6\%, 17)$$
$$= (\$15{,}000)(0.0954) + \$550 - (\$3000)(0.0354)$$
$$= \$1874.80 \quad (\$1900)$$

**Answer is C.**

**21.**
$$\sum M_A = 0$$

$$(11\text{ m})\left(50\ \frac{\text{N}}{\text{m}}\right)(6\text{ m}) + (8\text{ m})(100\text{ N}) + (8\text{ m})R_B = 0$$
$$R_B = -512.5\text{ N}$$
$$\sum F_x = 0$$
$$R_{Ax} = +512.5\text{ N}$$
$$\sum F_y = 0$$

$$-\left(50\ \frac{\text{N}}{\text{m}}\right)(6\text{ m}) - 100\text{ N} + R_{Ay} = 0$$
$$R_{Ay} = 400\text{ N}$$

**Answer is D.**

**22.** Member ED is a simply-supported beam.

$$\text{total load} = (6\text{ m})\left(50\ \frac{\text{N}}{\text{m}}\right) = 300\text{ N}$$

The reactions at F and D are

$$\frac{300\text{ N}}{2} = 150\text{ N}$$

The vertical load at point E is carried totally by member EF.

**Answer is C.**

**23.**

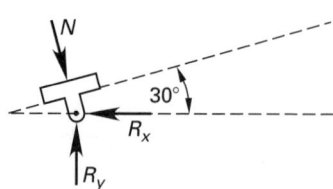

$$\sum M_A = 0$$
$$(1\text{ cm})N - (1\text{ cm} + 15\text{ cm})(30\text{ N}) = 0$$
$$N = \frac{(16\text{ cm})(30\text{ N})}{1\text{ cm}} = 480\text{ N}$$

$$R_y = N\cos 30° = (480\text{ N})(\cos 30°) = 415.7\text{ N}$$
$$p = \frac{R_y}{A_D} = \frac{415.7\text{ N}}{\dfrac{\pi\,(0.625\text{ cm})^2}{4}}$$
$$= 1355\text{ N/cm}^2$$
$$F_E = pA_E = \left(1355\ \frac{\text{N}}{\text{cm}^2}\right)\left(\frac{\pi\,(1.5\text{ cm})^2}{4}\right)$$
$$= 2394\text{ N}$$

**Answer is C.**

**24.**
$$\sum M_A = (6 \text{ m})(1000 \text{ N}) - (8 \text{ m})(500 \text{ N})$$
$$+ (12 \text{ m})(800 \text{ N})\cos 60°$$
$$- (16 \text{ m})(800 \text{ N})\sin 60° + (12 \text{ m})R_B$$
$$= 0$$
$$R_B = 357.1 \text{ N}$$

$$\sum F_x = 357.1 \text{ N} + R_{Ax} + 1000 \text{ N} + (800 \text{ N})(\cos 60°)$$
$$= 0$$
$$R_{Ax} = -1757.1 \text{ N}$$

$$\sum F_y = R_{Ay} - 500 \text{ N} - (800 \text{ N})(\sin 60°) = 0$$
$$R_{Ay} = 1192.8 \text{ N}$$

**Answer is A.**

**25.**
$$\sum F_y = -50 \text{ N} + R_{\text{wall}} = 0$$
$$\mathbf{R} = (50 \text{ N})\mathbf{j}$$

$$\sum M = \mathbf{M}_{\text{wall}} + (1 \text{ m})(50 \text{ N})\mathbf{k} + (2 \text{ m})(50 \text{ N})\mathbf{i} = 0$$
$$\mathbf{M}_{\text{wall}} = -(100 \text{ N·m})\mathbf{i} - (50 \text{ N·m})\mathbf{k}$$

**Answer is D.**

**26.** $\phi = \dfrac{TL}{JG}$

$$= \frac{(1 \text{ m})(50 \text{ N})(2 \text{ m})}{\frac{1}{2}\pi\left(\left(\frac{3 \text{ cm}}{2}\right)\left(\frac{1 \text{ m}}{100 \text{ cm}}\right)\right)^4 \left(79 \times 10^9 \ \frac{\text{N}}{\text{m}^2}\right)}$$

$$= 0.016 \text{ rad}$$

$$\phi = (0.016 \text{ rad})\left(\frac{360°}{2\pi \text{ rad}}\right) = 0.91°$$

**Answer is A.**

**27.** $y_{\text{max}} = \dfrac{FL^3}{3EI}$

$$= \frac{(50 \text{ N})(2 \text{ m})^3}{(3)\left(210 \times 10^9 \ \frac{\text{N}}{\text{m}^2}\right)\left(\frac{1}{4}\pi\left(\frac{0.03 \text{ m}}{2}\right)^4\right)}$$

$$= 0.016 \text{ m} \quad (1.6 \text{ cm})$$

**Answer is D.**

**28.**
$$\sigma_x = \frac{Mc}{I} = \frac{(50 \text{ N})(2 \text{ m})\left(\frac{0.03}{2} \text{ m}\right)}{\frac{1}{4}\pi\left(\frac{0.03}{2} \text{ m}\right)^4}$$

$$= 37.7 \times 10^6 \text{ N/m}^2 \quad (38 \text{ MPa})$$

**Answer is C.**

**29. Answer is D.**

**30.**
$$\text{percent } B_1 = 2.4\%$$
$$\text{percent } B_2 = 3.8\%$$

By the lever rule,

$$\text{percent solid} = \frac{3.8\% - 3.0\%}{3.8\% - 2.4\%} = 0.571 \quad (57\%)$$

**Answer is B.**

**31. Answer is B.**

**32.** Calculate the bar's cross-sectional area from the relationship between the mass, density, and volume.

$$\text{mass} = \rho V = \rho A L = 150 \text{ kg}$$
$$A = \frac{\text{mass}}{\rho L}$$
$$= \left(\frac{150 \text{ kg}}{\left(2500 \ \frac{\text{kg}}{\text{m}^3}\right)(4 \text{ m})}\right)\left(100 \ \frac{\text{cm}}{\text{m}}\right)^2$$
$$= 150 \text{ cm}^2$$

**Answer is D.**

**33.** A material is considered elastoplastic when the inelastic region of the stress-strain diagram is idealized as a straight line. Of the choices, metal is the only elastoplastic material. The other choices are all viscoelastic materials, which exhibit time dependent elastic strain.

**Answer is B.**

**34.** A material that has a high elastic modulus, high yield point, high elongation at fracture, and high ultimate strength is hard and tough. This is best depicted by the stress-strain diagram of option A.

**Answer is A.**

**35.** Low alloy steels are one of the most commonly used classes of structural steels, so option A is false. There are only two basic types of stainless steels: magnetic (martensitic) and non-magnetic (austenitic). Option B is false. The addition of small amounts of silicon to steel increases both the yield strength and tensile strength. Option C is false. The addition of small amounts of molybdenum to low-alloy steels makes it possible to harden and strengthen thick pieces of the metal by heat treatment.

**Answer is D.**

**36.** From the impulse-momentum theorem,

$$F = \dot{m}\Delta v = \rho v A v = v^2 A\rho$$

$v^2$ is found from

$$\rho gh = \frac{\rho v^2}{2}$$
$$v^2 = 2gh$$

Therefore,

$$h = \frac{v^2}{2g} = \frac{\dfrac{F}{A\rho}}{2g} = \frac{F}{2gA\rho}$$

$$= \frac{2.5 \text{ N}}{(2)\left(9.81 \dfrac{m}{s^2}\right)\pi\left(\dfrac{0.01}{2} \text{ m}\right)^2 \left(1000 \dfrac{kg}{m^3}\right)}$$

$$= 1.62 \text{ m} \quad (1.6 \text{ m})$$

**Answer is B.**

**37.**

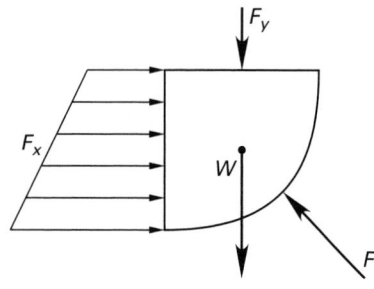

The average depth is

$$\overline{h} = \left(\tfrac{1}{2}\right)(h_1 + h_2) = \left(\tfrac{1}{2}\right)(4 \text{ m} + 7 \text{ m})$$
$$= 5.5 \text{ m}$$

$$\overline{p} = \rho g\overline{h} = \left(1000 \dfrac{kg}{m^3}\right)\left(9.81 \dfrac{m}{s^2}\right)(5.5 \text{ m})$$
$$= 53\,955 \text{ N/m}^2$$

$$F_x = \overline{p}A = \left(53\,955 \dfrac{N}{m^2}\right)(3 \text{ m})(1 \text{ m})$$
$$= 161\,865 \text{ N}$$

$$F_y = (3 \text{ m})(4 \text{ m})(1 \text{ m})\left(1000 \dfrac{kg}{m^3}\right)\left(9.81 \dfrac{m}{s^2}\right)$$
$$= 117\,720 \text{ N}$$

$$W = \left(\dfrac{\pi(3 \text{ m})^2}{4}\right)(1 \text{ m})\left(1000 \dfrac{kg}{m^3}\right)\left(9.81 \dfrac{m}{s^2}\right)$$
$$= 69\,342 \text{ N}$$

$$F = \sqrt{F_x^2 + (F_y + W)^2}$$
$$= \sqrt{(161\,865 \text{ N})^2 + (117\,720 \text{ N} + 69\,342 \text{ N})^2}$$
$$= 247\,300 \text{ N} \quad (250 \text{ kN})$$

**Answer is D.**

**38.**

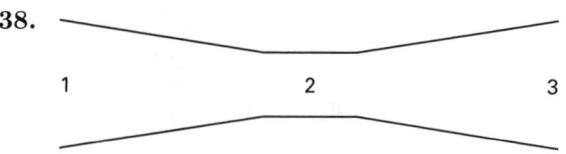

$$p_1 - p_{atm} = 172.4 \text{ kPa}$$
$$p_1 = 273.7 \text{ kPa}$$
$$v_1 A_1 = v_2 A_2$$
$$v_2 = v_1\left(\frac{A_1}{A_2}\right) = v_1\left(\frac{D_1}{D_2}\right)^2$$
$$p_1 + \frac{\rho v_1^2}{2} = p_2 + \frac{\rho v_2^2}{2}$$
$$p_2 = p_1 + \left(\frac{\rho v_1^2}{2}\right)\left(1 - \left(\frac{D_1}{D_2}\right)^4\right)$$
$$= 273.7 \text{ kPa}$$

$$p_2 = 273.7 \text{ kPa} + \left(\frac{\left(1000 \dfrac{kg}{m^3}\right)\left(7.62 \dfrac{m}{s}\right)^2}{2}\right)$$

$$\times \left(1 - \left(\frac{38.1 \text{ cm}}{21.24 \text{ cm}}\right)^4\right)\left(\frac{1 \text{ kPa}}{1000 \dfrac{N}{m^2}}\right)$$

$$= 2.15 \text{ kPa} \quad (2.2 \text{ kPa})$$

Cavitation is impending; $p_{vapor} = p_2$.

**Answer is A.**

**39.**

$$(N_{Fr})_{model} = (N_{Fr})_{ship}$$
$$\frac{v_{model}^2}{gL_{model}} = \frac{v_{ship}^2}{gL_{ship}}$$
$$v_{model} = \sqrt{\left(\frac{L_{model}}{L_{ship}}\right)v_{ship}^2}$$
$$= \sqrt{\left(\frac{1}{25}\right)(110 \text{ kph})^2}$$
$$= 22 \text{ kph}\left(0.278 \dfrac{\dfrac{m}{s}}{kph}\right)$$
$$= 6.11 \text{ m/s} \quad (6.1 \text{ m/s})$$

**Answer is B.**

**40.** The hydraulic radius is found by

$$R = \frac{D}{4} = \frac{0.5}{4} = 0.125 \text{ m}$$

The velocity of flow when the pipe is full is

$$v_{full} = \left(\frac{1.00}{n}\right) R^{2/3}\sqrt{S}$$

$$= \left(\frac{1.00}{0.015}\right)(0.125 \text{ m})^{2/3}\sqrt{0.001}$$

$$= 0.53 \text{ m/s}$$

The flow if the pipe is full is

$$Q_{full} = v_{full}A$$

$$= \left(0.53 \ \frac{\text{m}}{\text{s}}\right)\left(\frac{\pi}{4}\right)(0.5 \text{ m})^2$$

$$= 0.10 \text{ m}^3/\text{s}$$

$$\frac{Q}{Q_{full}} = \frac{0.07 \ \frac{\text{m}^3}{\text{s}}}{0.10 \ \frac{\text{m}^3}{\text{s}}} = 0.7$$

Using a graph of hydraulic elements for circular sewers, for $Q/Q_{full} = 0.7$, $v/v_{full} = 0.94$.

$$v = v_{full}\left(\frac{v}{v_{full}}\right)$$

$$= \left(0.53 \ \frac{\text{m}}{\text{s}}\right)(0.94)$$

$$= 0.50 \text{ m/s}$$

**Answer is A.**

**41.** The critical depth must be found first. (Note, the NCEES Reference Handbook uses $y_c$ instead of $d_c$, and $B$ instead of $w$.)

$$d_c^3 = \frac{Q^2}{gw^2}$$

$$d_c = \sqrt[3]{\frac{\left(14 \ \frac{\text{m}^3}{\text{s}}\right)^2}{\left(9.81 \ \frac{\text{m}}{\text{s}^2}\right)(6 \text{ m})^2}}$$

$$= 0.822 \text{ m}$$

The critical velocity is

$$v_c = \sqrt{gd_c}$$

$$= \sqrt{\left(9.81 \ \frac{\text{m}}{\text{s}^2}\right)(0.822 \text{ m})}$$

$$= 2.84 \text{ m/s} \quad (2.8 \text{ m/s})$$

**Answer is C.**

**42.** The area of the channel is found by the following equation, with $b$ as the width and $d$ as the depth. $\theta = 26.6°$, $\tan\theta = 1/2$.

$$A = \left(b + \frac{d}{\tan\theta}\right)d$$

$$= \left(5 \text{ m} + \frac{7 \text{ m}}{\frac{1}{2}}\right)(7 \text{ m})$$

$$= 133 \text{ m}^2$$

Use the continuity equation and solve for v.

$$Q = vA$$

$$v = \frac{Q}{A} = \frac{7 \ \frac{\text{m}^3}{\text{s}}}{133 \text{ m}^2}$$

$$= 0.0526 \text{ m/s} \quad (0.05 \text{ m/s})$$

**Answer is A.**

**43.** Use the equation for finding the velocity in a pitot-static gauge.

$$v = \sqrt{\frac{2gh(\rho_m - \rho)}{\rho}}$$

$$= \sqrt{\frac{(2)\left(9.81 \ \frac{\text{m}}{\text{s}^2}\right)(0.076 \text{ m}) \times \left(13\,580 \ \frac{\text{kg}}{\text{m}^3} - 1000 \ \frac{\text{kg}}{\text{m}^3}\right)}{1000 \ \frac{\text{kg}}{\text{m}^3}}}$$

$$= 4.33 \text{ m/s} \quad (4.3 \text{ m/s})$$

**Answer is D.**

**44.** Use the rational method equation, $Q = CIA$, to solve for peak flow of the system. The overland flow time is given for each parcel. The time for water to flow from the farthest corner to reach the detention pond is $t_c = 15 \text{ min} + 10 \text{ min} = 25 \text{ min}$.

The runoff coefficients are given for each area. Determine the composite $C$ value for the entire area.

$$C = \frac{(2000 \text{ m}^2)(0.35) + (4000 \text{ m}^2)(0.65)}{2000 \text{ m}^2 + 4000 \text{ m}^2}$$

$$= 0.55$$

The intensity after 25 minutes was given as 3.9 m/h. The total area is

$$4000 \text{ m}^2 + 2000 \text{ m}^2 = 6000 \text{ m}^2$$

The peak flow is found as follows.

$$Q = CIA$$

$$= (0.55)\left(3.9\ \frac{m}{h}\right)(6000\ m^2)\left(\frac{1\ h}{3600\ s}\right)$$

$$= 3.575\ m^3/s \quad (4.0\ m^3/s)$$

**Answer is B.**

**45.**    $C_{eq} = 20\ \mu F + 20\ \mu F = 40\ \mu F$

**Answer is D.**

**46.**    $i = C_{eq}\dfrac{dV}{dt} = (40 \times 10^{-6}\ F)\left(300\cos 60t\ \dfrac{V}{s}\right)$

$$= 0.012\cos 60t\ A$$

$$i_{eff} = \frac{i_{max}}{\sqrt{2}} = \frac{0.012\ A}{\sqrt{2}}$$

$$= 8.49 \times 10^{-3}\ A \quad (8.5\ mA)$$

**Answer is C.**

**47.**  For any purely capacitive circuit,

$$P_{dissipated} = 0\ mW$$

**Answer is A.**

**48.**  Shorting the voltage source and opening the current source gives

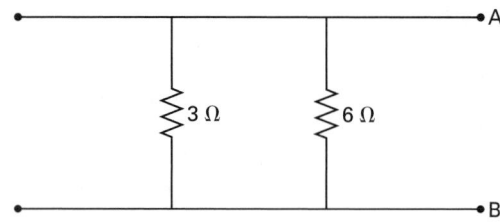

The equivalent Thevenin resistance is

$$R_{Th} = \frac{(3\ \Omega)(6\ \Omega)}{3\ \Omega + 6\ \Omega} = 2\ \Omega$$

Since the voltage source is across a resistor, the voltage drop across that resistor must be the applied voltage. Similarly, all current from the current source must go through the 23 Ω resistor, so the 23 Ω resistor is irrelevant in determining the terminal voltage. The equivalent circuit is

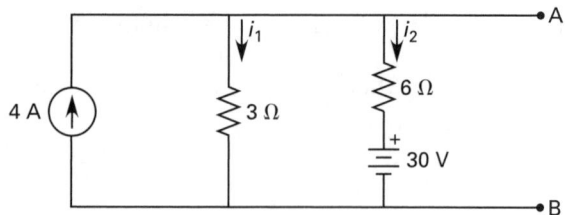

The two known equations are

$$i_1 + i_2 = 4\ A$$
$$(i_1)(3\ \Omega) = (i_2)(6\ \Omega) - 30\ V$$

Solving simultaneously,

$$i_2 = \frac{42}{9}\ A$$

$$i_1 = 4\ A - \frac{42}{9}\ A = -\frac{6}{9}\ A$$

The voltage drop across terminals A and B is

$$V_{Th} = (3\ \Omega)\left(-\frac{6}{9}\ A\right) = -2\ V$$

The Thevenin equivalent is

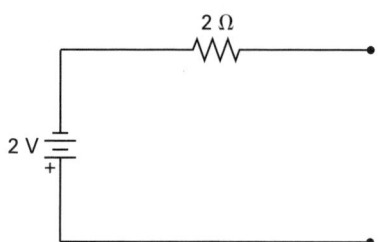

**Answer is B.**

**49.**    $\cos\phi_1 = 0.72$

$$\phi = \cos^{-1}(0.72) = 43.95°$$
$$\cos\phi_2 = 0.86$$
$$\phi_2 = \cos^{-1}(0.86) = 30.68°$$
$$P_{real} = P_{apparent}\cos\phi_1$$
$$= (5000\ kVA)(0.72) = 3600\ kVA$$
$$= 3.6 \times 10^6\ W$$
$$\Delta P_{reactive} = P_{real}(\tan\phi_1 - \tan\phi_2)$$
$$= (3.6 \times 10^6\ W)(\tan 43.95° - \tan 30.86°)$$
$$= 1.3 \times 10^6\ W$$
$$C = \frac{\Delta P_{reactive}}{2\pi f(V_{line,rms})^2} = \frac{1.3 \times 10^6\ W}{(2\pi)(60\ Hz)(220\ V)^2}$$
$$= 0.07\ F \quad (70\ mF)$$

**Answer is C.**

**50.**  $X_L = 2\pi fL = (2\pi)(60 \text{ Hz})(0.35 \text{ H})$

$= 131.9 \ \Omega$

$X_C = \dfrac{1}{2\pi fC} = \dfrac{1}{(2\pi)(60 \text{ Hz})(40 \times 10^{-6} \text{ F})}$

$= 66.3 \ \Omega$

The total reactance is

$X_L - X_C = 131.9 \ \Omega - 66.3 \ \Omega$

$= 65.6 \ \Omega \quad (66 \ \Omega)$

**Answer is A.**

**51.**  The magnetic permeabilities of steels vary greatly. The saturation depends on the type of steel used in the poles.

**Answer is A.**

**52.**  $W = \left(\dfrac{p_1 V_1}{k - 1}\right)\left(1 - \left(\dfrac{p_2}{p_1}\right)^{\frac{k-1}{k}}\right)$

$= \left(-\dfrac{(101\,325 \text{ Pa})(0.566 \text{ m}^3)}{1.15 - 1}\right)$

$\times \left(1 - \left(\dfrac{334\,500 \text{ Pa}}{101\,325 \text{ Pa}}\right)^{\frac{1.15-1}{1.15}}\right)$

$= 64\,400 \text{ J} \quad (64.4 \text{ kJ})$

**Answer is D.**

**53.**

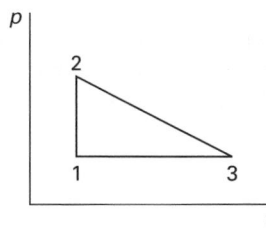 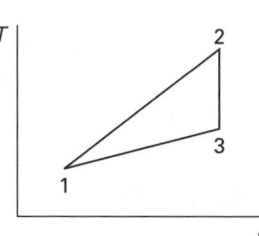

$V_1 = V_2$

$p_2 = \left(\dfrac{T_2}{T_1}\right)p_1 = \left(\dfrac{200°\text{C} + 273}{20°\text{C} + 273}\right)(101\,325 \text{ kPa})$

$= 163\,572 \text{ Pa} \quad (163.6 \text{ kPa})$

$s_2 = s_3$

$T_3 = T_2\left(\dfrac{p_3}{p_2}\right)^{\frac{k-1}{k}}$

$= (200°\text{C} + 273)\left(\dfrac{101.325 \text{ kPa}}{163.6 \text{ kPa}}\right)^{\frac{1.4-1}{1.4}}$

$= 412.5\text{K}$

$= 139°\text{C}$

$Q_{in} = Q_{12} = mc_v(T_2 - T_1)$

$= (1 \text{ kg})\left(718 \ \dfrac{\text{J}}{\text{kg·K}}\right)(200°\text{C} - 20°\text{C})$

$= 129 \text{ kJ}$

$Q_{out} = -Q_{31} = mc_p(T_1 - T_3)$

$= (1 \text{ kg})\left(1000 \ \dfrac{\text{J}}{\text{kg·K}}\right)(20°\text{C} - 139°\text{C})$

$= -119 \text{ kJ}$

$\eta_{Th} = \dfrac{W_{net}}{Q_{in}} = \dfrac{Q_{in} - Q_{out}}{Q_{in}} = \dfrac{129 \text{ kJ} - 119 \text{ kJ}}{129 \ \dfrac{\text{kg·K}}{\text{kJ}}}$

$= 0.0775 \quad (7.8\%)$

**Answer is A.**

**54.**  $m_1 = \dfrac{p_1 V}{RT_1} = \dfrac{\left(3.45 \times 10^6 \ \dfrac{\text{Pa}}{\text{MPa}}\right)(20 \text{ m}^3)}{\left(273 \ \dfrac{\text{J}}{\text{kg·K}}\right)(20°\text{C} + 273)}$

$= 862.6 \text{ kg}$

$T_2 = T_1\left(\dfrac{p_2}{p_1}\right)^{\frac{k-1}{k}}$

$= (20°\text{C} + 273)\left(\dfrac{1.725 \text{ MPa}}{3.45 \text{ MPa}}\right)^{\frac{1.4-1}{1.4}}$

$= 240.4\text{K}$

$m_2 = \dfrac{p_2 V_2}{RT_2} = \dfrac{\left(1.725 \times 10^6 \ \dfrac{\text{Pa}}{\text{MPa}}\right)(20 \text{ m}^3)}{\left(273 \ \dfrac{\text{J}}{\text{kg·K}}\right)(240.4\text{K})}$

$= 525.7 \text{ kg}$

$\Delta m = m_1 - m_2 = 862.6 \text{ kg} - 525.7 \text{ kg}$

$= 337 \text{ kg}$

**Answer is A.**

**55.**  $n = \dfrac{pV}{R^* T}$

$= \dfrac{(1 \text{ atm})(500 \text{ L})}{\left(0.08206 \ \dfrac{\text{atm·L}}{\text{mol·K}}\right)(20.56°\text{C} + 273)}$

$= 20.75 \text{ mol}$

The available energy is

$q = \eta n(\text{heating value})$

$= (0.90)(20.75 \text{ mol})\left(802 \ \dfrac{\text{kJ}}{\text{mol}}\right)$

$= 14\,977 \text{ kJ}$

Energy is required to raise the ice temperature, melt the ice, and raise the water temperature. For 1.0 kg of ice, the required energy is

$$(c_{p,\text{ice}})(\Delta T) + \text{heat of fusion} + (c_{p,\text{water}})(\Delta T)$$

$$\left(2.11 \ \frac{\text{kJ}}{\text{kg} \cdot {}^\circ\text{C}}\right)(0 - (-17.8^\circ\text{C})) + 333 \ \frac{\text{kJ}}{\text{kg}}$$

$$+ \left(4.18 \ \frac{\text{kJ}}{\text{kg} \cdot {}^\circ\text{C}}\right)(37.8^\circ\text{C} - 0^\circ\text{C})$$

$$= 528.6 \ \text{kJ/kg}$$

The mass of ice is

$$m = \frac{14\,977 \ \text{kJ}}{528.6 \ \dfrac{\text{kJ}}{\text{kg}}}$$

$$= 28.3 \ \text{kg} \quad (28 \ \text{kg})$$

**Answer is C.**

**56.** The piston will move until $P_A = P_B$. Therefore, $P_{Bf} = 5$ atm. For a perfect gas, $pV^k = \text{constant}$.

$$P_{Bi}V_{Bi}^k = P_{Bf}V_{Bf}^k$$

$$V_{Bf} = \left(\frac{P_{Bi}}{P_{Bf}}\right)^{\frac{1}{k}} V_{Bi} = \left(\frac{1 \ \text{atm}}{5 \ \text{atm}}\right)^{\frac{1}{1.4}} (1 \ \text{m}^3)$$

$$= 0.317 \ \text{m}^3 \quad (0.32 \ \text{m}^3)$$

**Answer is D.**

**57.**
$$W = VU = mC_v\Delta T$$
$$= mC_v(T_f - T_i)$$

Assume isentropic compression.

$$T_f = T_i\left(\frac{P_f}{P_i}\right)^{\frac{k-1}{k}} = (300\text{K})\left(\frac{5 \ \text{atm}}{1 \ \text{atm}}\right)^{\frac{1.4-1}{1.4}}$$

$$= 475\text{K}$$

$$m = \frac{P_iV_i}{RT_i} = \frac{(1 \ \text{atm})\left(101\,325 \ \dfrac{\text{Pa}}{\text{atm}}\right)(1 \ \text{m}^3)}{\left(8.315 \ \dfrac{\text{J}}{\text{mol} \cdot \text{K}}\right)(300\text{K})}$$

$$= 40.62 \ \text{mol}$$

$$W = mC_v(T_f - T_i)$$

$$= (40.62 \ \text{mol})\left(20 \ \frac{\text{J}}{\text{mol} \cdot \text{K}}\right)$$

$$\times (475\text{K} - 300\text{K})\left(\frac{1 \ \text{kJ}}{1000 \ \text{J}}\right)$$

$$= 142.2 \ \text{kJ} \quad (140 \ \text{kJ})$$

**Answer is D.**

**58.**
$$Q = mc_v\Delta T$$

$$\Delta T = \frac{Q}{mc_v}$$

$$Q = (200 \ \text{Btu})\left(1.055 \ \frac{\text{kJ}}{\text{Btu}}\right)$$

$$= 211 \ \text{kJ}$$

$$\Delta T = \frac{211 \ \text{kJ}}{(2 \ \text{kg})\left(4.2 \ \dfrac{\text{kJ}}{\text{kg} \cdot \text{K}}\right)}$$

$$= 25.12\text{K} \quad (25\text{K})$$

**Answer is B.**

**59.** $R = \dfrac{1.98 \ \dfrac{\text{cal}}{\text{gmol} \cdot \text{K}}}{32 \ \dfrac{\text{g}}{\text{gmol}}}$

$$= 0.0619 \ \frac{\text{cal}}{\text{g} \cdot \text{K}}$$

$$T = 20^\circ\text{C} + 273^\circ = 293\text{K}$$

$$W = -\int_{V_1}^{V_2} p\,dV$$

$$= -mRT\ln\frac{V_2}{V_1}$$

$$= -(40 \ \text{g})\left(0.0619 \ \frac{\text{cal}}{\text{g} \cdot \text{K}}\right)(293\text{K})\ln\left(\frac{0.60}{1}\right)$$

$$= -370.59 \ \text{cal} \quad (370 \ \text{cal})$$

**Answer is C.**

**60.** Maximum work output will be obtained in a reversible process. The difference between the maximum and the actual work output is known as the process irreversibility.

**Answer is D.**